The Rebirth of Mankind
Homo Evolutis

By: Trent Goodbaudy

Published in Portland, Oregon by PDXdzyn. PDXdzyn is a trademark of Trent Goodbaudy. PDXdzyn titles may be purchased in bulk for educational, business, fund-raising, or sales promotional use. For information, please send email to info@pdxdzyn.com.

Printed in the United States of America.

ISBN: 1470108844
ISBN-13: 978-1470108847

Trent Goodbaudy

DEDICATION

This book is dedicated to my daughter, Crystalynn Rose.

CONTENTS

"By responsible use of science, technology, and other rational means we shall eventually manage to become posthuman."

—Nick Bostrom

Trent Goodbaudy

ACKNOWLEDGMENTS

A very special thanks to the following; Aaron Franz, Sophia Smallstrom, Clifford Carnicom, Jan Smith, and all those who seek truth.

INTRODUCTION

A basic activity in military intelligence involves gathering information from many sources to arrive at an understanding of a larger picture, in order for the commander to make decisions appropriate for the conditions. This is what I have attempted to do in this book, and the implications are both shocking and frightening.

Technology has made some amazing strides in the last few decades; from advances in computers and electronic devices, to nanotechnology, genetics, informatics, synthetic biology, geo-engineering and much more.

Man can now even control the weather! For example, in 2010 scientists created 50 rainstorms in the desert (Abu Dhabi's Al Ain region), and in a confidential company video, the founder of the Swiss company in charge of the project, Metro Systems International, boasted of success. Helmut Fluhrer said: "*We have achieved a number of rainfalls.*" It is believed to be the first time the system has produced rain from clear skies, according to Fluhrer. In the past, China and other countries have used chemicals for cloud-seeding to both induce and prevent rain falling. In June 2010, Metro Systems built five ionizing sites each with 20 emitters which can send trillions of cloud-forming ions into the atmosphere. Over four summer months the emitters were switched on when the required atmospheric level of humidity reached 30 per cent or more. While the country's weather experts predicted no clouds or rain in the Al Ain region, rain fell on FIFTY-TWO occasions. The project was monitored by the Max Planck Institute for Meteorology, one of the world's major centers for atmospheric physics. Professor Hartmut Grassl, a former institute director, said: "*There are many applications. One is getting water into a dry area.*" The scientists have been working secretly for United Arab Emirates president Sheikh Khalifa bin Zayed Al Nahyan.

These kinds of technological advances will be sold to us as necessary, and absolutely beneficial, they will also be released to us incrementally... in versions. Just think about how much people are in love with their iToys right now, and how long have we even had mobile wireless devices? What if we determine that all these radio frequencies emitting from these wireless devices are causing irreparable damage to our bodies at a cellular level, or even simply disrupting the natural energy flow in our bodies and perhaps making cancers and other abnormalities more likely? We need to be very careful, or we can potentially cause the extinction or permanent modification of the entire human race. Just think, if we are releasing particles into the air (supposedly to make it rain) and modifying and even synthesizing nature at an atomic scale, what kinds of safeguards do we need to have in place before we commence implementing these new technologies? What if the plants and other life on the planet absorb the metallic ionizers that are being released into the atmosphere? What if what is actually being sprayed on us is way more serious than we know?

I believe that the Human race is getting ready to endure some significant tribulation in the near future. The absolute worst thing about what we are currently going through, and what we will continue to go through is the rapid breakneck pace with which advancement and culmination of several new advancements will transpire, and how safeguards will be overlooked, and accidents will happen. Even worse, what if we were forced to change at a molecular level... through environmental factors that were out of our control?

That which we term "reality" - as understood mentally by our thinking processes - is currently being altered and changed into different expressions to allow complete control of our personal reality - a complete absolution of that which has been normally operative in a divinely ordained, natural frequency. We must not allow these corporations, governments and private research funding from figures like Bill Gates to succeed without making sure mankind is being protected. Instead of providing safeguards and protection however, the powers that be are using the global masses' ignorance of this very advanced technology to their advantage.

The elevated scientific methods described in this book have been suppressed for many years and only those very few people in power have been privy to this information. The confidentiality with which these operations and projects operate under is substantiation that they have also likely breached ethics policies at the very least. I am curious as to what makes the research in these breakthrough fields so important that we need to risk our very natural existence to advance in them. I think that the government and other powers that be are scared of losing their iron-grip of fear-based control over the people., and they are working hard to maintain this control through military applications of these technologies, and even manipulating the fundamental genetic structure of the Homo Sapien species to design and create an entirely new advanced and highly evolved species; "Homo Evolutis".

I will attempt to explore many of these emerging technologies in the subsequent chapters within this book, but please understand that I am not a professional by any means in the technologies that follow. I do try and help break down concepts into easy-to-digest morsels to aid in incorporation and comprehension of the technologies and concepts involved. This book is simply a result of incredible amounts of diligent research, making amazing connections, and exposing alarming revelations. I believe that it will help us all to understand exactly where humanity is being guided, and why.

Do you lack energy lately? Having interrupted sleep? Are you already experiencing symptoms of one of the many emerging diseases? Are you already diagnosed as having a disease? Do you eat GMO foods grown by the industrial scale farms processed by large packagers? Do you live surrounded by Wi-Fi now becoming immersing everyone and spreading out and expanding all the time in developed countries? Do you use your cell phone daily? Do you use a microwave oven daily? Do you watch the 11:00 o'clock news on high definition television? Do you live anywhere near an industrial scale wind generator? Do you have the "smart meter" put on yet? Do you have a GPS in your vehicle? Do you work in front of a computer screen all day? Have you been vaccinated for anything? Did you know about the corporate ties between Microsoft, Google, Skype, Facebook with a logo of HAARP? If not you may well already use Facebook or Twitter with your personal life permanently profiled and recorded for datamining. Do you want to potentially be turned into a zombie, a robotic, controlled entity walking around in an apathetic state, or do you want to live life as a vital, vibrant human being... as the Divine Creator intended for you? We are all living on a very small spaceship called Earth, the problem is... we have no escape shuttles. We, as united inhabitants of this planet, can and must turn the tide.

"Awareness is the Cure. "

1 AN "AGE OF TRANSITIONS"

New Years Eve, 1999 - The long awaited new millennium was finally upon us, and although it was a time of great celebration; the night was met with much apprehension about the threat of the Y2K bug, and of course even back then the United States was on alert from threats of terrorism as well. As the midnight hour passed and nothing catastrophic happens; everyone feels an even greater cause for celebration as we enter a new millennium full of hope for the future. The entire world passes a mile marker of human existence, and all of mankind enters a new age.

On this New Years eve of the dawn of the new millennium, the current U.S. President Clinton makes a speech. In this speech he says things about the future that, looking back now, had a very serious forecast for humanity. He says, "*Tonight we celebrate the change of centuries, the dawning of a new millennium. We celebrate the future, imagining an even more remarkable 21st century... So we Americans must not fear change, instead let us welcome it, embrace it, and create it... Such a triumph will require great efforts from us all. It will require us to stand against the forces of hatred and bigotry, terror and destruction. It will require us to make further breakthroughs in science and technology. To cure diseases, heal broken bodies, lengthen life and unlock secrets, from global warming; to the black holes in the universe.*"

Although Y2K turned out fine, there was a definite sense of disappointment, the coming of the new millennium seemed anti-climactic. There was no discernable difference between the years 1999 and 2000. As the year passed, this feeling only grew stronger, as the presidential election featured arguably two of the most boring candidates in history; George W. Bush, and Al Gore.

There was some excitement though, as an extremely close election ended in controversy over who had won. When the dust had settled, the title of "Commander in Chief" went to the second president in the history of the United States with the last name of Bush.

Then it was back to business as usual, with the most boring president in history. CBS news reported "Our CBS News tally shows President George W. Bush made 9 visits to his Texas ranch, spending all or part of 69 days there during his first year in office. Mr. Bush also made 25 visits to Camp David in 2001 totaling 78 days. And he spent a four-day weekend at his folks' place in Kennebunkport, Maine that year." And even the Comedy Central Network has a video segment available online from an August 9, 2001 "Daily Show with Jon Stewart" where the president was interviewed about world events from his "Mobile tactical command center" (Otherwise known as a golf cart). Another video available online shows a clip of the president saying; "*I'm working on some initiatives ... we're a ... you'll see, I*

mean there will be ... I've got some ... there will be some decisions ... we will be announcing them as time goes on."

Then on September 11, 2001 something happened that would change the world forever. That morning we watched in horror and disbelief, as reality resembled the fiction of a big-budget Hollywood disaster film. The nation was in shock.

How did this happen? Who flew those planes? What are we going to do? These were the obvious questions which demanded answers. And we didn't have to wait very long for those answers; nineteen Muslim extremists had hijacked airliners with box cutters. They targeted major US landmarks to get their message across, which we were told was *"We hate America, and it's freedom"*. These men were led by Osama Bin Laden; the mastermind behind the anti-American, radical Islamic group known as Al-Qaida.

Luckily, American troops had already been amassed in and around Afghanistan, the country where Bin Laden was hiding in a cave. As the military went to work in Afghanistan, we the people were asked to make some serious changes at home. The "War on Terror" had begun, and with it came the expectation that certain liberties would have to be sacrificed, in the name of security. We found ourselves living in a "post 9/11" world. As the world changed all around us, a fanatical form of patriotism offered condolence to those mourning the loss of the now distant "good old days". Boredom, was a thing of the past, and changes came faster than they ever had before.

While most people were slapping "red, white, and blue" stickers on their bumpers, the United States government was busy holding a conference concerned with changes so large, that they promised to alter human nature itself. 21st century goals were discussed in preparation for what would come to be known as "The age of transitions".

Convergent Technologies

The title "Age of Transitions" was coined by Newt Gingrich, in his introduction to the National Science Foundation, and Department of Commerce sponsored workshop titled "Converging Technologies For Improving Human Performance" on N.B.I.C. (nano, bio, info, cogno) technologies. The workshop featured a wide range of participants, from governmental and private institutions, such as; N.A.S.A, the Department of Defense, M.I.T., Carnegie Mellon University, University of California

Berkeley, Defense Advanced Research Projects Agency (D.A.R.P.A.), Hewlett Packard, Stanford University, and many more.

It was a chance for experts in the field of Nano scale, Biology, Information, and Cognitive technologies to discuss their visions for the future, alongside government officials. The goals discussed for the future were nothing less than promethean, with the key goal stated as "*converging technologies focused on enhancing human performance*", which in turn would lead to "*a more efficient societal structure*".

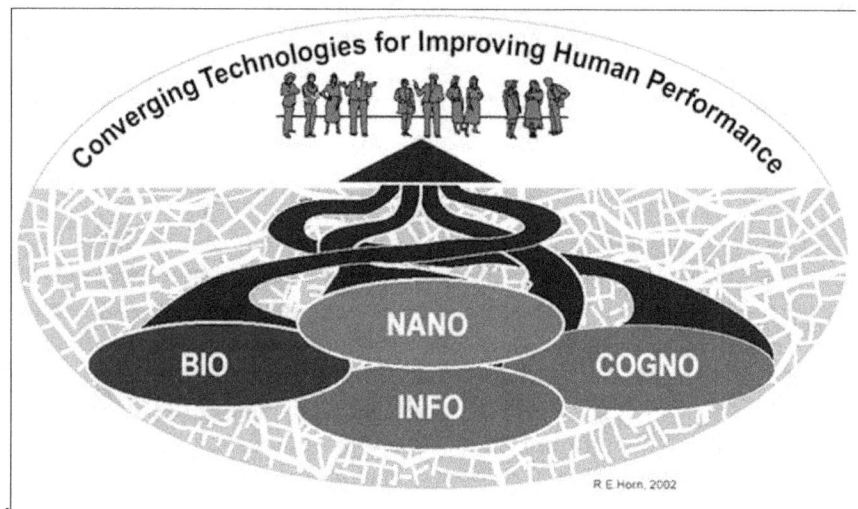

From Roco and Bainbridge (ed.). (2002). 'Converging Technologies for Improving Human Performance: Nanotechnology, Biotechnology, Information Technology and Cognitive Science', NSF/DOC-sponsored report, Virginia.

Indeed, technological convergence was given as the answer to all of the world's, now infamous "global problems". It promised to bring upon a new renaissance of human development. It was the hope of integrating humanity with nature, to save the earth.

Visions laid out included: robotics, cybernetics, artificial intelligence, life extension, brain enhancement, virtual reality, genetic engineering, and even teleportation. Enhancing human "performance" would require merging human biology with technology. Brain machine interfaces (or BMIs), would allow the control of machinery with the brain itself.

Implantable brain chips would also be able to store information and enhance cognitive function. The ultimate human-machine symbiosis could be to download a copy of a person's brain into a supercomputer. This would allow someone to effectively live forever within a computer-generated, virtual simulation.

More subtle concepts such as using virtual reality in classrooms were also discussed. And of course, the military implications of convergence were quite obvious. As cybernetic enhancement of human performance is inevitable. Achieving these visions requires the decoding and understanding of complex systems; the most important complex system being the human brain. After all, it is the driving force behind human performance. Through the use of bioinformatics, functions of the brain could be understood.

Bioinformatics is the process of collecting data from a biological system; in order to understand how that system works. The next step in the process would involve biomimetics, or the mimicking of those same biological systems. Using this process could enable the development of artificial intelligence.

A stated military goal for artificial intelligence, is the creation of uninhabited combat vehicles *"removing the pilot would result in a more combat-agile aircraft"*. These machines would also have the ability to maintain themselves. The use of new materials created with nanotech would enable lighter, stronger, high tech solutions.

Of course the "super-soldier" was also mentioned at this conference. All sorts of new techniques; from pharmaceuticals, to robotic exoskeletons could make this vision come true. It's important to realize that this report clearly states that the cybernetic enhancement of human performance is inevitable; *"Inevitably, the cybernetic enhancement of human performance is sneaking up on society."*

With Newt Gingrich proclaiming; "Those countries that ignore these patterns of change will fall further behind and find themselves weaker, poorer, and more vulnerable than their wiser, more change-oriented neighbors." Mr. Gingrich Conservatively calls for a tripling of the National Science Foundation's budget. He also mentions George Bush's approval of a $604,000,000 increase towards the "nano" budget. Convergence is the priority area of importance in implementing the great promise of a new day for the 21st century.

One group stands above all others in applauding this funding for convergence. They are known as the "World Transhumanist Association", W.T.A., or transhumanists. Most prominent among which; are professors, philosophers, scientists, and celebrities.

The transhumanists see a world of problems just begging to be solved with converging technology.

WTA co-founder Professor Nick Bostrom states;

> "Sometimes we don't see a problem because either, it's too familiar, or it's too big. The first is "death", death is a BIG problem. If you look at the statistics, the odds are not very favorable to us. So far, most people who have lived, have also died.
>
> Existential risk, the second BIG problem. Existential risk is a threat to human survival or to the long term potential of our species.
>
> The third BIG problem is that life isn't usually as wonderful as it could be. I think that's a big, big problem. There are just those moments that have experienced where life was fantastic, and you wonder, why can't it be like that all the time.
>
> Suppose we fixed these things."

The Singularity

The transhumanist "Golden Age" will kick off an event known as "The Singularity". The singularity will occur at the point which artificial intelligence surpasses the capabilities of the human brain.

Speaking of our current concept and state of robotic technology;

> "Is this really the best, we can dream of? Is this the best we can do? Or is it possible to find something a little more ... inspiring? If we want to achieve this; what in the world would have to change? And this is the answer; we would have to change, not just the world around us, but we ourselves. Not just the way we think about the world, but the way we are. Our very biology, human nature would have to change. So, we could think of

adding on, different new sensory capacities and mental faculties." - Prof. Nick Bostrom.

I honestly wonder what he truly meant when he said that "... *our very biology* ..." would have to change. That in itself is a frightening concept in my opinion.

"When you go out to the late twenty-twenties, everybody ... almost everybody will have some amount of non-biological intelligence inside their brains. It is going to happen in this very gradual way. By introducing nonbiological intelligence that gradually becomes more and more sophisticated with new versions. You get to the twenty-forties, and the nonbiological, the machine portion of our intelligence is going to be vastly more powerful than the biological portion. The biological intelligence will be trivial at that point, and ultimately that is where the action is." - Inventor and author, Ray Kurzwell

"My personal [view] is that we are the information, the processing going on in our minds. Which means that an 'uploaded' mind would be the same person. A computer copy of me, would be me." - Anders Sandberg Ph.D

"There will be cyborgs ... I see that as inevitable, but I don't think it's going to solve our problem, it will just make it worse. Because imagine that your neighbor starts going 'cyborgy', starts adding components, and suddenly your next door neighbor is capable of learning a human language in seconds. Then you would get a real split in human capacities because these cyborgs are no longer human in effect ... sort of superhuman." - Professor Dr. Hugo de Garis

"Now for the first time in human history, in the information revolution. We can begin to become masters of intelligence. This mastery will offer us unparalleled freedom and opportunities. It has the potential to enrich our lives more than anything we've seen before." - Dr. Michio Kaku

"All revolutions have winners and losers. This revolution is no exception, but I would say the big losers are the people who say they don't want to get involved. They are going to discover that

being a little bit out of touch will have some unpleasant consequences."

"They are going to get a lot of people who do want to upgrade themselves, there is no question about that. And there will be commercial interests, and political interests supporting those groups. There's a lot of money to be made here and a lot of power. Not just in a military sense but in an everyday sense, in terms of who gets jobs and who doesn't." - Professor Kevin Warwick

The Posthuman

From cyborgs, with very long life-spans, to downloading consciousness itself into a machine. Transhumanists say it is impossible to predict exactly what a "posthuman" will be. But it will indeed be better than human. Such lofty promises are understandably being embraced by many people, in search of a better world for everyone. So called "techno-progressives", wish to see this technology developed, but also want to make sure that it is equitably distributed throughout all strata of society.

Techno-progressives are rivaled by various people, including Bioconservatives. Bioconservatives are those people opposed to the creation of posthumans. Surprisingly, also by experts within the fields of convergence itself. Marvin Minsky, the originator of artificial neural networks, and the co-founder of the Artificial Intelligence lab at M.I.T. has said; *"Ordinary citizens wouldn't know what to do with eternal life... the masses don't have any clear-cut goals or purpose."*

No matter how you look at it, the singularity is being promised as *the* great solution to our 21st century global problems. Thanks to such early work as the human genome project, we will be able to decode DNA itself.

Through the use of applied genetics, science will be able to improve the human race.

What most people don't realize, is that this concept is not new. It is in fact a re-packaging of what was once called "eugenics".

2 EUGENICS

"Eugenics is the study of the agencies under social control that may improve or impair the racial qualities of future generations either physically or mentally." ~ Sir Francis Galton, 1904

~

The practice of selective breeding and domestication by humans of both plants and animals began many thousands of years ago. From the domestication of the dog in East Asia around 15,000 BC, and wild grasses 'encouraged' to be made into heartier foodstuffs dating back to 10,000 BC in Syria.

Wild animals have been domesticated; both as food and as servants to man for a very long time. So, why wouldn't it be beyond man to think that it can collectively guide the progression of its own species as well?

The notion of segregating people considered unfit to reproduce dates back to antiquity. There is support of this practice in the form of evidence of ritual sacrifice of people with unusual bone growth patterns in the European

Upper Paleolithic region (Old Stone Age, 26,000 - 8,000 BC). Research made by Vincenzo Formicola (University of Pisa, Italy) points to a significant occurrence of multiple burials, commonly connected to simultaneous deaths caused by natural disasters or diseases. But a detailed analysis showed that some of the multiple burials could have been selective. The skeletons in these graves present a sex and age correlation and in some of the most relevant sites skeletons belong to severely deformed individuals which apparently suffered from congenital malformations, like dwarfism or bone bowing.

Even within the Old Testament is described the Amalekites – a supposedly depraved group that God condemned to death.

Concerns about environmental influences that might damage heredity – leading to ill health, early death, insanity, and defective offspring – were formalized in the early 1700s as degeneracy theory. Degeneracy theory maintained a strong scientific following until late in the 19th century.

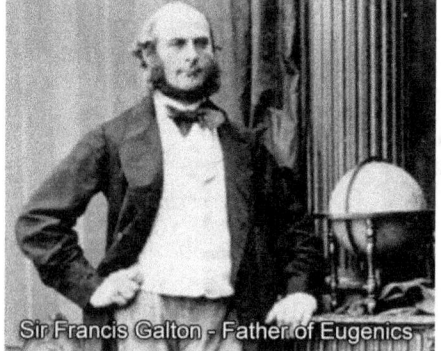
Sir Francis Galton - Father of Eugenics

The term eugenics comes from the Greek roots for "good" and "generation" or "origin" and was first used to refer to the "science" of heredity and good breeding in about 1883 by Sir Francis Galton.

Within 20 years of the first reference of "eugenics" in 1883, the word was widely used by scientists who had rediscovered the work of Gregor Mendel. Mendel had meticulously recorded the results of cross-breeding pea plants, and found a very regular statistical pattern for features like height and color. This introduced the concept of genes, opening the field of genetics to a frenzied century of research. One path of genetic research branched off into the shadows of social theory, and in the first quarter of the twentieth century became immensely popular as eugenics. It was presented as a mathematical science that could be used to predict the traits and behaviors of humans, and in a perfect world, to control human breeding so that people with the best genes would reproduce and thus improve the species. It was an optimistic school of thought with a profound faith in the powers of Science.

Masturbation, then called onanism, was presented in medical schools as the first biological theory of the cause of degeneracy. Fear of degeneracy through masturbation led Harry Clay Sharp, a prison physician in Jeffersonville, Indiana, to carry out vasectomies on prisoners beginning in 1899. The advocacy of Sharp and his medical colleagues, culminated in an Indiana law mandating compulsory sterilization of "degenerates." Enacted in 1907, this was the first eugenic sterilization law in the United States.

Sir Francis Galton, perceived eugenics as a moral philosophy to improve humanity by encouraging the ablest and healthiest people to have more children. The Galtonian ideal of eugenics is usually termed positive eugenics. Negative eugenics, on the other hand, advocated culling the least able from the breeding population to preserve humanity's fitness. The eugenics movements in the United States, Germany, and Scandinavia favored the negative approach.

By the mid-19th century most scientists believed bad environments caused degenerate heredity. Benedict Morel's work extended the causes of degeneracy to some legitimate agents – including poisoning by mercury, ergot, and other toxic substances in the environment. The sociologist Richard Dugdale believed that good environments could transform degenerates into worthy citizens within three generations. This position was a backdrop to his very influential study on The Jukes (1877), a degenerate family of paupers and petty criminals in Ulster County, New York. The inheritance of acquired (environmental) characters was challenged in the 1880s by August Weismann, whose theory of the germ plasm convinced most scientists that changes in body tissue (the soma) had little or no effect on reproductive tissue (the germ plasm). At the beginning of the 20th century, Weismann's views were absorbed by degeneracy theorists who embraced negative eugenics as their favored model.

Adherents of the new field of genetics were ambivalent about eugenics. Most basic scientists – including William Bateson in Great Britain, and Thomas Hunt Morgan in the United States – shunned eugenics as vulgar and an unproductive field for research. However, Bateson's and Morgan's contributions to basic genetics were quickly absorbed by eugenicists, who took interest in Mendelian analysis of pedigrees of humans, plants, and animals. Many eugenicists had some type of agricultural background.

Charles Davenport and Harry Laughlin, who together ran the Eugenics Record Office, were introduced through their shared interest in chicken breeding. Both also were active in Eugenics Section of the American

Breeder's Association (ABA). Davenport's book, Eugenics: The Science of Human Improvement through Better Breeding, had a distinct agricultural flavor, and his affiliation with the ABA was included under his name on the title page. Agricultural genetics also provided the favored model for negative eugenics: human populations, like agricultural breeds and varieties, had to be culled of their least productive members, with only the healthiest specimens used for breeding.

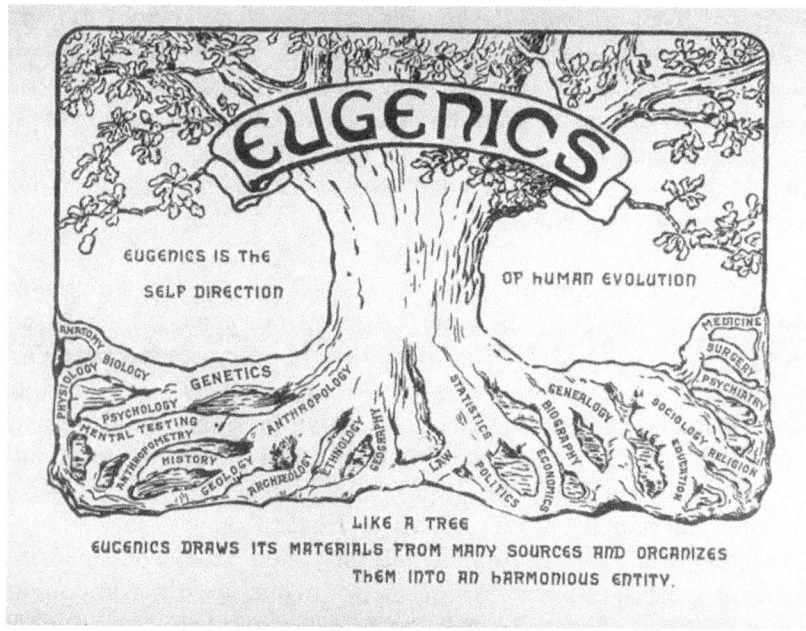

2nd International Congress of Eugenics held in 1921.

Evolutionary models of natural selection and dysgenic (bad) hereditary practices in society also contributed to eugenic theory. For example, there was fear that highly intelligent people would have smaller families (about 2 children), while the allegedly degenerate elements of society were having larger families of four to eight children. Public welfare might also play a role in allowing less fit people to survive and reproduce, further upsetting the natural selection of fitter people.

When Charles Darwin published his groundbreaking theory of Natural Selection in 1859, it was received by the public with considerable displeasure. Although the esteemed naturalist had been kind enough to explain his theory using mounds of logic and evidence, he lacked the good manners to incorporate the readers' preconceived notions of the universe. Nevertheless, many men of science were drawn to the elegant hypothesis, and they found it pregnant with intriguing corollaries. One of these was a phenomenon Darwin referred to as *artificial selection*: the centuries-old process of selectively breeding domestic animals to magnify desirable traits. This, he explained, was the same mechanism as natural selection, merely accelerated by human influence.

Eugenics was born as a scientific curiosity in the Victorian age. In 1863, Sir Francis Galton, a half-cousin of Charles Darwin, theorized that if talented people only married other talented people, the result would be measurably better offspring. At the turn of the last century, Galton's ideas were imported into the United States just as Gregor Mendel's principles of heredity were rediscovered. American eugenic advocates believed with religious fervor that the same Mendelian concepts determining the color and size of peas, corn and cattle also governed the social and intellectual character of man.

In 1865, Galton pried the lid from yet another worm-can with the publication of his article entitled "Hereditary Talent and Character." In this essay, the gentleman-scientist suggested that one could apply the principle of artificial selection to humans just as one could in domestic animals, thereby exaggerating desirable human traits over several generations. This scientific philosophy would come to be known as *eugenics*, and over the subsequent years its seemingly sensible insights gained approval worldwide. In an effort to curtail the genetic pollution created by "inferior" genes, some governments even enacted laws authorizing the forcible sterilization of the "*insane, idiotic, imbecile, feebleminded or epileptic,*" as well as individuals with criminal or promiscuous inclinations. Ultimately hundreds of thousands of people were forced or coerced into sterilization worldwide, over 65,000 of them in the country which pioneered the eugenic effort: The United States of America.

In an America demographically reeling from immigration upheaval and torn by post-Reconstruction chaos, race conflict was everywhere in the early twentieth century. Elitists, utopians and so-called "progressives" fused their smoldering race fears and class bias with their desire to make a better world. They reinvented Galton's eugenics into a repressive and racist ideology. The intent: populate the earth with vastly more of their own socio-economic and biological kind--and less or none of everyone else.

From the beginning, Sir Francis Galton and his league of extraordinary eugenicists were concerned that the human race was facing an inevitable decline. They worried that advances in medicine were too successful in improving the survival and reproduction of weak individuals, thereby working at odds with natural evolution.

Darwin himself expressed some concern regarding such *negative selection*:

"[We] do our utmost to check the process of elimination. We build asylums for the imbecile, the maimed and the sick; we institute poor-laws; and our medical men exert their utmost skill to save the life of every one to the last moment. [...] Thus the weak members of civilized societies propagate their kind. No one who has attended to the breeding of domestic animals will doubt that this must be highly injurious to the race of man. [...] Nor could we check our sympathy, even at the urging of hard reason, without deterioration in the noblest part of our nature."

Eugenicists argued that "defectives" should be prevented from breeding, through custody in asylums or compulsory sterilization. Most doctors probably felt that sterilization was a more humane way of dealing with people who could not help themselves.

Vasectomy and tubal ligation were favored methods, because they did not alter the physiological and psychological contribution of the reproductive organs.

Sterilization allowed the convicted criminal or mental patient to participate in society, rather than being institutionalized at public expense. Sterilization was not viewed as a punishment because these doctors believed (erroneously) that the social failure of "unfit" people was due to an irreversibly degenerate germ plasm.

The grand plan was to literally wipe away the reproductive capability of those deemed weak and inferior--the so-called "unfit." The eugenicists hoped to neutralize the viability of 10 percent of the population at a sweep, until none were left except themselves.

Eighteen solutions were explored in a Carnegie-supported 1911 "Preliminary Report of the Committee of the Eugenic Section of the American Breeder's Association to Study and to Report on the Best Practical Means for Cutting Off the Defective Germ-Plasm in the Human Population." Point eight was euthanasia.

The most commonly suggested method of eugenicide in America was a "lethal chamber" or public locally operated gas chambers. In 1918, Popenoe, the Army venereal disease specialist during World War I, co-wrote the widely used textbook, Applied Eugenics, which argued, "*From an historical point of view, the first method which presents itself is execution... Its value in keeping up the standard of the race should not be underestimated.*" Applied Eugenics also devoted a chapter to "Lethal Selection," which operated "*through the destruction of the individual by some adverse feature of the environment, such as excessive cold, or bacteria, or by bodily deficiency.*"

The Greatest Sin of all is total IGNORANCE of the most important subject in the life of every man and woman—SEX.

Away With False Modesty!

Let us face the facts of sex fearlessly and frankly, sincerely and scientifically. Let us tear the veil of shame and mystery from sex and build the future of the race on a new knowledge of all the facts of sex as they are laid bare in plain, daring but wholesome words, and frank pictures in the huge new library of Sex Knowledge.

"MODERN EUGENICS"

544 Pages of SECRETS

Everything a Married Woman Should Know—

How to hold a husband
How to have perfect children
How to preserve youth
Warding off other women
Keeping yourself attractive
Why husbands tire of wives
Dreadful diseases due to ignorance
Diseases of women
Babies and birth control
Twilight sleep—easy childbirth
How babies are conceived
Diseases of children
Family health guide
Change of life—hygiene
Why children die young
Inherited traits and diseases
What will you tell your growing girl?
The mystery of twins
Hundreds of valuable remedies

Secrets for Men—

Mistakes of early marriages
Secrets of fascination
Joys of perfect mating
How to make women love you
Bringing up healthy children
Fevers and contagious diseases

Accidents and emergencies
Hygiene in the home
Limitation of offspring
The sexual embrace
Warning to young men
Secrets of greater delight
Dangerous diseases
Secrets of sex attraction
Hygienic precautions
Anatomy and physiology
The reproductive organs
What every woman wants
Education of the family
Sex health and prevention

**Girls—
Don't Marry**

before you know all this—

The dangers of petting
How to be a vamp
How to manage the honeymoon
What liberties to allow a lover
Secrets of the wedding night
Beauty diets and baths
Do you know—
How to attract desirable men
How to manage men
How to know if he loves you
How to acquire bodily grace and beauty
How to beautify face, hands, hair, teeth and feet
How to acquire charm
How to dress attractively
Intimate personal hygiene
How to pick a husband

Eugenic breeders believed American society was not ready to implement an organized lethal solution. But many mental institutions and doctors practiced improvised medical lethality and passive euthanasia on their own. One institution in Lincoln, Illinois fed its incoming patients milk from tubercular cows believing a eugenically strong individual would be immune. Thirty to forty percent annual death rates resulted at Lincoln. Some doctors practiced passive eugenicide one newborn infant at a time. Other doctors at mental institutions engaged in lethal neglect.

Nonetheless, with eugenicide marginalized, the main solution for eugenicists was the rapid expansion of forced segregation and sterilization, as well as more marriage restrictions. California led the nation, performing nearly all sterilization procedures with little or no due process. In its first twenty-five years of eugenic legislation, California sterilized 9,782 individuals, mostly women. Many were classified as "bad girls," diagnosed as "passionate," "oversexed" or "sexually wayward." At Sonoma, some women were sterilized because of what was deemed an abnormally large clitoris or labia.

As a cautionary measure, many US states enacted laws as early as 1896 prohibiting marriage to anyone who was "epileptic, imbecile or feeble-minded". But in 1907, eugenics truly passed the threshold from hypothesis into practice when the state of Indiana erected legislation based upon the notion that socially undesirable traits are hereditary:

"…it shall be compulsory for each and every institution in the state, entrusted with the care of confirmed criminals, idiots, rapists and imbeciles, to appoint upon its staff, in addition to the regular institutional physician, two (2) skilled surgeons of recognized ability, whose duty it shall be, in conjunction with the chief physician of the institution, to examine the mental and physical condition of such inmates as are recommended by the institutional physician and board of managers. If, in the judgment of this committee of experts and the board of managers, procreation is inadvisable and there is no probability of improvement of the mental condition of the inmate, it shall be lawful for the surgeons to perform such operation for the prevention of procreation as shall be decided safest and most effective."

1907 INDIANA EUGENICS LAW
• • •
By late 1800s, Indiana authorities believed criminality, mental problems, and pauperism were hereditary. Various laws were enacted based on this belief. In 1907, Governor J. Frank Hanly approved first state eugenics law making sterilization mandatory for certain individuals in state custody. Sterilizations halted 1909 by Governor Thomas R. Marshall.
(Continued on other side)

Indiana eugenics marker.

Although this particular law was later overturned, it is widely considered to be the world's first eugenic legislation. The sterilization of imbeciles was put into practice, often without informing the patient of the nature of the procedure. Similar laws were soon passed elsewhere in the US, many of which withstood the legal gauntlet and remained in force for decades.

Meanwhile the founders of the newly-formed Eugenics Record Office in New York began to amass hundreds of thousands of family pedigrees for genetic research. The organization publicly endorsed eugenic practices, and lobbied for state sterilization acts and immigration restrictions. The group also spread their vision of genetic superiority by sponsoring a series of "Fitter Families" contests which were held at state fairs throughout the US. Alongside the state's portliest pigs, swiftest horses, and most majestic vegetables, American families were judged for their quality of breeding. Entrants' pedigrees were reviewed, their bodies examined, and their mental capacity measured. The families found to be most genetically fit were awarded a silver trophy, and any contestant scoring a B+ or higher was awarded a bronze medal bearing the inscription, "Yea, I have a goodly heritage."

The eugenics movement took another swerve for the sinister in 1924 when the state of Virginia enacted a matched set of eugenics laws: The Sterilization Act, a variation of the same sterilization legislation being passed throughout the US; and the Racial Integrity Act, a law which felonized marriage between white persons and non-whites.

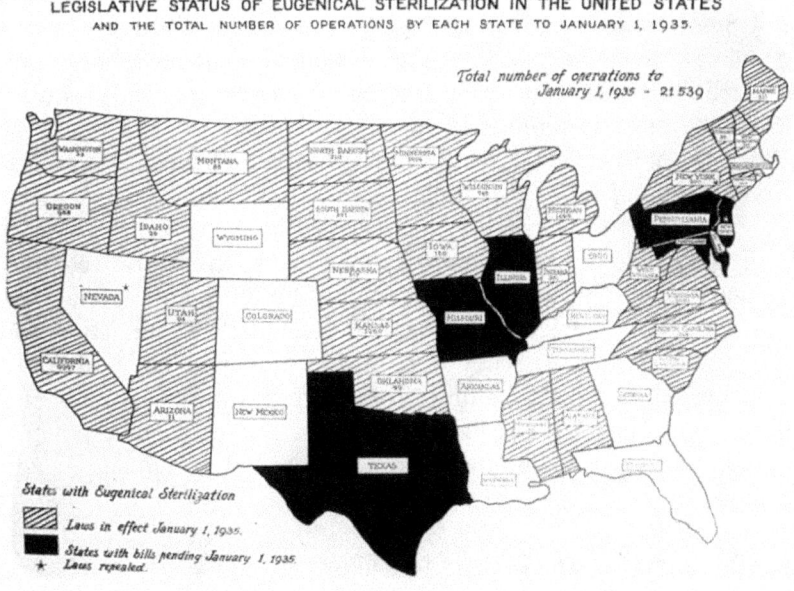

LEGISLATIVE STATUS OF EUGENICAL STERILIZATION IN THE UNITED STATES AND THE TOTAL NUMBER OF OPERATIONS BY EACH STATE TO JANUARY 1, 1935.

In September of the same year, this shiny new legislation was challenged by a patient at the Virginia State Colony for Epileptics and Feebleminded. Eighteen-year-old Carrie Buck − child to a promiscuous mother, and mother to an illegitimate child − refused her mandatory sterilization and a legal challenge was arranged on her behalf. A series of appeals ultimately brought the *Buck v. Bell* case before the Supreme Court of the United States. The Supreme Court's ruling was delivered by Justice Oliver Wendell Holmes, Jr.: "*It is better for all the world, if instead of waiting to execute degenerate offspring for crime, or to let them starve for their imbecility, society can prevent those who are manifestly unfit from continuing their kind. The principle that sustains compulsory vaccination is broad enough to cover cutting the Fallopian tubes…Three generations of imbeciles are enough.*"

With the apparent vindication of these myopic eugenics laws, sterilization procedures were ordered by the thousands. Carrie Buck and her daughter Vivian were among them. It was later discovered that Carrie had been become pregnant with Vivian after being raped by her foster parents' nephew, and that her commitment into the Colony had been a gambit to preserve the family's reputation. It seems that Carrie was neither feebleminded nor promiscuous, she was merely inconvenient.

These sorts of *negative eugenics* policies enjoyed widespread adoption in the US and Canada throughout the 1920s and 30s, with some lawmakers contemplating plans to make welfare and unemployment relief contingent upon sterilization. In the years leading up to the Second World War, however, the eugenic philosophy received the endorsement of the Nazis, and their "racial hygiene" atrocities rapidly dragged the eugenic philosophy from public favor.

Eugenics would have been no more than just so much bizarre parlor talk had it not been for extensive financing by corporate philanthropies, specifically the Carnegie Institution, the Rockefeller Foundation and the Harriman railroad fortune. They were all in league with some of America's most respected scientists hailing from such prestigious universities as Stanford, Yale, Harvard, and Princeton. These academicians espoused race theory and race science, and then faked and twisted data to serve eugenics' racist aims.

Stanford president David Starr Jordan originated the notion of "*race and blood*" in his 1902 racial epistle "Blood of a Nation," in which the university scholar declared that human qualities and conditions such as talent and poverty were passed through the blood.

In 1904, the Carnegie Institution established a laboratory complex at Cold Spring Harbor on Long Island that stockpiled millions of index cards on ordinary Americans, as researchers carefully plotted the removal of families, bloodlines and whole peoples. From Cold Spring Harbor, eugenics advocates agitated in the legislatures of America, as well as the nation's social service agencies and associations.

The Harriman railroad fortune paid local charities, such as the New York Bureau of Industries and Immigration, to seek out Jewish, Italian and other immigrants in New York and other crowded cities and subject them to deportation, trumped up confinement or forced sterilization.

The Rockefeller Foundation helped found the German eugenics program and even funded the program that Josef Mengele worked in before he went to Auschwitz.

When Nazi leaders were put on trial for war crimes, they cited the United States as the inspiration for the 450,000 forced sterilizations they conducted. The eugenic laws in the US remained in force, however, and sterilization programs continued quietly for many years thereafter. One by one the state laws were repealed, and by 1963 virtually all US states had dismantled their sterilization legislation– but not before 65,000 or so imbeciles, criminals, and fornicators were surgically expelled from the gene pool. As for the legal precedent of *Buck v. Bell*, it has yet to be officially overruled.

Even with the shifts in public opinion, concerns regarding the decline of the species still remained. It was believed that certain undesirable diseases could be reduced or eliminated from humanity through well-informed mate selection, including such maladies as hypertension, obesity, diabetes, heart disease, muscular dystrophy, cystic fibrosis, hemophilia, and certain types of cancer. In an effort to improve general quality of life, some scientists hypothesized that the ideal way to save humanity would be for healthy and attractive women to breed with men of science. Unfortunately, no orgy of intellectuals ensued.

The breeding behaviors of humans remains of utmost interest to geneticists today. In Israel, the *Dor Yeshorim* organization was founded to provide genetic screenings for couples considering marriage. If it is discovered that both the man and woman carry the recessive gene for *Tay-Sachs* disease– a genetic defect which causes a slow, painful death within a child's first five years– the couple are advised against marrying. The same process screens for several other hereditary diseases which are common among Jews, and owing to this eugenic guidance, the number of affected individuals has been reduced considerably. A similar screening system has been successful in nearly eradicating the disease *thalassemia* on the island of Cyprus. Such applications align with the original vision of eugenics before it became distorted by misguided minds: voluntary, altruistic, and based upon scientifically measurable criteria. Unfortunately the imperfections in screening methods have occasionally led to bizarre "wrongful life" lawsuits, where disabled individuals seek compensation for their unprevented afflictions.

It seems that the eugenic philosophy of *intelligent evolution* is inseparable from humanity's future– and we have only just begun to open the massive ethical worm-cans. Historian Daniel Kevles from Yale University suggests that eugenics is akin to the conservation of natural resources; both can be practiced horribly so as to abuse individual rights, but both can be practiced wisely for the betterment of society. There is no doubt that the forced sterilizations in the name of eugenics were an indefensible trespass upon the rights of individuals; what will be forced upon us next time? Will we even know before it is too late?

3 A BRIEF OVERVIEW AND
HISTORY OF GENETICS

Genetics in simple terms is just the study of genes, and this chapter attempts to explain what genes do and how they work. Living organisms are able to pass genetic information to each new generation, and this information manifests in traits, or a feature of a living thing. A trait can determine differences in physical appearance. For example, hair color, eye color, height, and blood type are all determined by different traits. A combination of genetic information is passed down to us from our parents, and we inherit a combination of characteristics from them that ultimately makes each of us unique.

The way we interact with our environment can also affect our traits and even our genes, but this can become complicated. The chances of someone dying of heart disease or cancer seems to be dependent on the environment and lifestyle in combination with the information contained within the genes.

Your body is made up of approximately one-hundred-trillion (100,000,000,000,000) cells. There are many different types of cells that perform many different functions within our bodies constantly, each of them operating independently while also providing a valuable service to the organ it is a part of. Some of the more common types of cells are skin cells, blood cells, fat cells, and nerve cells. Each of these types of cells perform a function that is valuable to the organism or complex system that we call our body.

Inside almost every cell is something called a nucleus, which contains 99.9% of your genes. There is also mitochondria in each cell that contains a small amount of genetic information. Genes are small parts of a molecule called Deoxyribonucleic acid (or DNA). We each have a total of around 20,000 genes within our DNA (give or take). If you were to extract your DNA from the nucleus of a cell and then uncoil it into one strand, it would measure around 6 feet long! All of this genetic information is coiled up into a single cell nucleus, and each cell contains this information that is the foundation of life for all of us.

The chemical structure of DNA allows it to act as a chemical code that carries genetic information. All of the information it contains is found on a long string of four chemical "letters": A for adenine, C for cytosine, G for guanine and T for thymine. These chemical compounds are the building blocks of DNA.

Each gene is composed of a stretch of DNA containing myriad of these chemical letters. In effect, DNA is the set of blueprints for the proteins that carry out our biochemical functions.

The following scientists have been considered major players in the research of DNA; Francis Crick, James Watson and Maurice Wilkins. Some important highlights from 19th century to the present 21st century era include the following:

1869 - Friedrich Miescher discovers a new acid in the nuclei of white blood cells, later identified as DNA.

1928 - Fred Griffith uncovers the existence of a "transforming principle" that transfers genetic instructions between bacteria.

1944 - Oswald Avery announces that genes are made up of DNA, not protein as previously believed and that DNA is in fact Griffith's "Transforming Principle."

1953 - Francis Crick and James Watson publish the "double helix" structure of DNA

1974 - First experiments in genetic engineering, transferring DNA between different organisms, are conducted at Stanford University.

2003 - The Human Genome Project, launched under James Watson in 1990, deciphers the human genetic code.

The symbol of the double helix is now a famous structure of DNA, a model that uses just four chemical letters to signify a body, a life.

Genetic Mutation

In molecular biology and genetics, mutations are permanent changes within the DNA sequence of a cell's genome or the DNA or RNA sequence of a virus. They can be defined as sudden and spontaneous changes in the cell. Gene mutations can be inherited from a parent, or they can happen during a person's lifetime due to lifestyle, environment, or other causes.

Some of the other causes of mutations include, radiation, viruses, mutagenic chemicals, as well as errors that occur during meiosis or DNA replication. They can also be induced by the organism itself, by cellular processes such as hyper-mutation.

Mutations can range in size from a single DNA building block (DNA base) to a large segment of a chromosome. While organisms can normally carry genes with detrimental mutations and are likely to survive with the mutations, it should be noted that severely detrimental mutations could either kill an organism or prevent it from passing on its genes. Mutations that kill the organism are called "lethals".

Mutation can result in several different types of changes in sequences; these can either have no effect, alter the product of a gene, cause an active gene to go dormant and vice-versa, or prevent the gene from functioning properly or completely.

Studies of the fruit fly (Drosophila melanogaster) suggest that if a mutation changes a protein that is produced via a gene, this will probably be harmful, with about 70 percent of these mutations having damaging effects, and the remainder being either neutral or weakly beneficial. Due to the damaging effects that mutations can have on genes, organisms have mechanisms such as DNA repair to prevent mutations.

Some regions of DNA control other genes, determining when and where these other genes are turned "on" or "off". Mutations in these parts of the genome can substantially change the way the organism is built. The difference between a mutation to a control gene and a mutation to a less powerful gene is a bit like the difference between sending an instruction to the trumpet player in an orchestra versus sending it to the orchestra's conductor. The impact of changing the conductors behavior will have a much larger and more coordinated effect than changing the behavior of an individual orchestra member. Similarly, a mutation in a gene "conductor" can cause a cascade of effects in the behavior of genes under its control, and subsequently can also have a negative impact on other systems and processes within the body. Imagine someone that has webbed feet, or in an extreme situation mutations to a control gene could result in Siamese twins.

Acquired (or *somatic*) mutations take place in the non-reproductive cells at some time during a person's life. These changes can be caused by environmental factors such as ultraviolet radiation from the sun, or can occur if a mistake is made as DNA copies itself during cell division. These type of mutations cannot be passed on to the next generation.

Mutations that occur in either an egg or sperm cell or those that occur just after fertilization, are called new (de novo) mutations. De novo mutations may explain genetic disorders in which an affected child has a mutation in every cell, but has no family history of the disorder.

Inherited mutations that are passed from parent to child are called hereditary mutations. This type of mutation is present throughout a person's life in virtually every cell in the body. These are the mutations that matter to evolutionists and eugenicists.

Some genetic changes are very rare; others are common in the population. Genetic changes that occur in more than 1 percent of the population are called polymorphisms. They are common enough to be considered a normal variation in the DNA. Polymorphisms are responsible for many of the normal differences between people such as eye color, hair color, and blood type. Although many polymorphisms have no negative effects on a person's health, some of these variations may influence the risk of developing certain disorders.

Genetic Evolution

Evolution is any change across successive generations in the heritable characteristics of biological populations. Evolutionary processes give rise to diversity at every level of biological organization, including species, individual organisms and molecules such as DNA and proteins.

Evolutionary genetics is the field of studies that resulted from the integration of genetics and Darwinian evolution. Genetic evolution can also be thought of as genetic mutation or genetic drift that affects a large percentage of the population and also be considered beneficial. This term is used by eugenicists to explain beneficial mutations, or a method to create a scientific argument for analyzing, guiding, and promoting the beneficial mutations. This could also make a strong argument for intentionally creating "beneficial" mutations. The question arises; to whom would these mutations benefit? A benefit to one individual could be considered a detriment to another. Who would a constructed mutation be beneficial to? The answer would be that typically the planner of the "beneficial" mutation would have the most to gain.

Charles Darwin was the first to gain widespread acceptance by formulating a scientific argument for the theory of evolution by means of natural selection. Evolution by natural selection is a process that is inferred from three facts about populations: 1) more offspring are produced than can possibly survive, 2) traits vary among individuals, leading to differential rates of survival and reproduction, and 3) trait differences are heritable. Thus, when members of a population die they are replaced by the progeny of parents that were better adapted to survive and reproduce in the environment in which natural selection took place. This process creates and preserves traits that are seemingly fitted for the functional roles they perform. Natural selection is the only known cause of adaptation, but not the only known cause of evolution. Nonadaptive causes of evolution include mutation and genetic drift.

In the early 20th century, genetics was integrated with Darwin's theory of evolution by natural selection through the discipline of population genetics. The importance of natural selection as a cause of evolution was accepted into other branches of biology.

Moreover, previously held notions about evolution, such as orthogenesis and "progress" became obsolete. Scientists continue to study various aspects of evolution by forming and testing hypotheses, constructing scientific theories, using observational data, and performing experiments in both the field and the laboratory. Biologists agree that descent with modification is one of the most reliably established facts in science. Discoveries in evolutionary biology have made a significant impact not just within the traditional branches of biology, but also in other academic disciplines (e.g., anthropology and psychology) and on society at large.

Evolutionary genetics has also been called the 'modern synthesis' (*Huxley* 1942), achieved through the theoretical works of R. A. Fisher, S. Wright, and J. B. S. Haldane and the conceptual works and influential writings of J. Huxley, T. Dobzhansky, and H.J. Muller.

Modern Synthesis is a theory about how evolution works at the level of genes, phenotypes, and populations whereas Darwinism was concerned mainly with organisms, speciation and individuals. This is a major paradigm shift and those who fail to appreciate it find themselves out of step with the thinking of evolutionary biologists. Many instances of such confusion can be in the popular press, and in other tentacles of the mainstream media.

Many people do not understand current ideas about evolution. The following is a brief summary of the modern consensus among evolutionary biologists.

The idea that life on Earth has evolved was widely discussed in Europe in the late 1700's and the early part of the 1800's. In 1859 Charles Darwin supplied a mechanism, namely natural selection, that could explain how evolution occurs. Darwin's theory of natural selection helped to convince most people that life has evolved and this point has not been seriously challenged in the past one hundred and fifty years.

It is important to note that Darwin's book "The Origin of Species by Means of Natural Selection" did two things. It summarized all of the evidence in favor of the idea that all organisms have descended with modification from a common ancestor, and thus built a strong case for evolution. In addition Darwin advocated natural selection as a mechanism of evolution. Biologists no longer question whether evolution has occurred or is occurring. That part of Darwin's book is now considered to be so overwhelmingly demonstrated that it is often referred to as the FACT of evolution. However, the MECHANISM of evolution is still debated.

We have learned much since Darwin's time and it is no longer appropriate to claim that evolutionary biologists believe that Darwin's theory of Natural Selection is the best theory of the mechanism of evolution. I can understand why this point may not be appreciated by the average non-scientist because natural selection is easy to understand at a superficial level. It has been widely promoted in the popular press and the image of "survival of the fittest" is too powerful and too convenient.

During the first part of this century the incorporation of genetics and population biology into studies of evolution led to a Neo-Darwinian theory of evolution that recognized the importance of mutation and variation within a population. Natural selection then became a process that altered the frequency of genes in a population and this defined evolution. This point of view held sway for many decades but more recently the classic Neo-Darwinian view has been replaced by a new concept which includes several other mechanisms in addition to natural selection. Current ideas on evolution are usually referred to as the Modern Synthesis which is described by Futuyma;

"The major tenets of the evolutionary synthesis, then, were that populations contain genetic variation that arises by random (ie. not adaptively directed) mutation and recombination; that populations evolve by changes in gene frequency brought about by random genetic drift, gene flow, and especially natural selection; that most adaptive genetic variants have individually slight phenotypic effects so that phenotypic changes are gradual (although some alleles with discrete effects may be advantageous, as in certain color polymorphisms); that diversification comes about by speciation, which normally entails the gradual evolution of reproductive isolation among populations; and that these processes, continued for sufficiently long, give rise to changes of such great magnitude as to warrant the designation of higher taxonomic levels (genera, families, and so forth)."
- Futuyma, D.J. in Evolutionary Biology, Sinauer Associates, 1986; p.12

This description would be incomprehensible to Darwin since he was unaware of genes and genetic drift. The modern theory of the mechanism of evolution differs from Darwinism in three important respects:

It recognizes several mechanisms of evolution in addition to natural selection. One of these, random genetic drift, may be as important as natural selection.

It recognizes that characteristics are inherited as discrete entities called genes. Variation within a population is due to the presence of multiple alleles of a gene.

It postulates that speciation is due to the gradual accumulation of small genetic changes. This is equivalent to saying that macroevolution is simply a lot of microevolution.

This field attempts to account for evolution in terms of changes in gene and genotype frequencies within populations and the processes that convert the variation with populations into more or less permanent variation between species. In this view, four evolutionary forces (mutation, random genetic drift, natural selection, and gene flow) acting within and among populations cause micro-evolutionary change and these processes are sufficient to account for macro-evolutionary patterns, which arise in the longer term from the collective action of these forces. That is, given very long periods of time, the micro-evolutionary forces will eventually give rise to the macro-evolutionary patterns that characterize the higher taxonomic groups. Thus, the central challenge of Evolutionary Genetics is to describe how the evolutionary forces shape the patterns of biodiversity observed in nature.

The force of mutation (historically) has been the ultimate source of new genetic variation within populations. Although most mutations are neutral with no harmful effect on fitness, some mutations have a small, positive effect on fitness and these variants are the raw materials for gradualistic adaptive evolution. Within finite populations, random genetic drift and natural selection affect the mutational variation. Natural selection is the only evolutionary force which can produce adaptation, the fit between organism and environment, or conserve genetic states over very long periods of time in the face of the dispersive forces of mutation and drift. The force of migration or gene flow has effects on genetic variation that are the opposite of those caused by random genetic drift. Migration limits the genetic divergence of populations and so impedes the process of speciation. The effect of each of these evolutionary forces on genetic variation within and among populations has been developed in great detail in the mathematical theory of population genetics founded on the seminal works of Fisher, Wright, and Haldane.

Some scientists continue to refer to modern thought in evolution as Neo-Darwinian. In some cases these scientists do not understand that the field has changed but in other cases they are referring to what can be called the Modern Synthesis, only they have retained the old name.

Among the evolutionary forces, natural selection has long been privileged in evolutionary studies because of its crucial role in adaptation. Ecological genetics is the study of evolutionary processes, especially adaptation by natural selection, in an ecological context in order to account for phenotypic patterns observed in nature. Where population genetics tends toward a branch of applied mathematics founded on Mendelian axioms, often with minimal contact with data, ecological genetics is grounded in the reciprocal interaction between mathematical theory and empirical observations from field and laboratory.

Genetic Modification

Genetic modification (GM), genetic manipulation (GM) and genetic engineering (GE) all refer to the same thing – the use of modern biotechnology techniques to change the genes of an organism, such as a plant or animal. A genetically modified organism (GMO) is a plant, animal or other organism that has been changed using genetic modification. Most do not like to think of ourselves as something that can be genetically modified. A person would be wrong however, if they thought that it was something that was currently not being studied and implemented.

It is an interesting fact that most proposals of improving the human body in transhumanistic discussions are mainly based upon bionic and chemical enhancements, while overlooking the potential of genetic engineering. In part this may be due to the fact that most methods of changing the genome are most efficient only on very small groups of cells or in the embryo. This means that these methods will mainly work on our children, not on ourselves, something which has made many transhumanists turn to other methods. However, genetic engineering has obviously great potential to transform living beings, it is already a viable technology (unlike bionics) and gene therapy is advancing fast. Perhaps most important, and controversial, is the fact that this method will not only change a single individual, but also affect all of his/her/its offspring. This will give us the ability to once and for all eliminate certain genes or add new ones.

Traditional breeding of plants and animals aims to tailor the plant or animal for a certain application. For example, a new crop variety might be bred that is more drought tolerant or resistant to a certain disease. The process of traditional breeding involves finding individuals with favorable traits and crossing them with each other – with the aim that the progeny of the cross will have the favorable traits from both parents. In reality the progeny have a mix of traits, both good and bad from their parents, and it takes a number of breeding cycles to eliminate the negative traits and build on the positive.

The final new plant variety or breed of animal will hopefully just have the desired traits, which it will have inherited from its ancestors along with the associated genes for those traits. Traditional breeding is a way of harnessing the genetic resources of an organism by breeding out unwanted genes and breeding in desirable genes.

GM breeding is used because it can change the genes of an organism in ways not possible through traditional breeding techniques providing opportunities for new plant varieties and animal breeds.

GM includes using genes from one organism and inserting them into another. For example, insect resistant GM cotton uses a gene from a naturally occurring soil bacterium to provide it with built-in insect protection. The use of insect resistant GM cotton has reduced pesticide use over 80 per cent in Australia.

However, GM does not necessarily mean that a gene from another organism has to be used to create the GMO. GM can mean that the organism's own genes are changed.

For example, gene silencing turns down the activity of certain genes already within an organism, such as in oilseed crops where it is being used to turn down the production of unhealthy oils. GM is also used for purely research purposes, for example, to discover genes.

There are many modifications that can be made to humans which are possible according to what we know, and reasonable extrapolations of current technology. This means that most of these enhancements currently will work only on the molecular level and not in the lesser understood areas of morphogenesis or other high-level functions.

Simple Modifications

The following modifications are mainly concerned with removing undesirable parts of the genome and changes between different naturally occurring alleles.

Removal of genetic defects / Removal of genetic diseases

These two categories overlap to a great extent. They include mutations of important genes, omissions or accidental overlaps in the genome. Many diseases seem to have a genetic factor, for example Alzenheimer's disease, glaucoma, certain forms of obesity, retinal detachment, diabetes II and cancer, and there are many more genes that weaken the body or make some diseases more likely.

Removal of undesirable traits

Of course, what is considered undesirable is often a highly speculative matter, and many negative traits are linked to positive traits in a complex manner. For example, the "novelty gene", which induces "novelty-seeking behavior" (i.e. adventure-seeker) will under the right circumstances make a person a dynamic neophile, but could also increase the risk of drug addiction (Reward Deficiency Syndrome); Dyslexia might be linked with visual thinking.

One solution to this is perhaps to inhibit the expression of the undesired genes, but provide a mechanism to remove inhibition if the owner of the gene so desires (this could possibly be accomplished through the creation of artificial genetic switches, which could be controlled using artificial hormones, although it is much more complex than simply removing the gene). Unfortunately this will not have any effect on genes responsible for the formation of organs or the body, since they are used only during development and then lie dormant.

Also note that removal of a gene linked to an undesirable trait may not completely prevent it's expression, since many of these traits are linked both to several genes and environmental factors. Some possible genetic traits which could be removed (or added) are:

Alcoholism

Drug Abuse

Schizophrenia

Cancer

Diabetes

Mano-Depressivity

Extreme Aggression

Wisdom Teeth

Cosmetic Modifications - These include, but are certainly not limited to:

Hair color, style and growth

Eye color

Skin color

Build

Adding Desirable Traits - There are some alleles that appear to promote health or other useful traits. For example, the D allele of the ACE gene increases endurance slightly and certain alleles of Apo-lipoprotein E protect against Alzheimer's disease.

Other changes - There is no doubt that there are many other possible alterations in appearance which could be developed. This is mainly dependent on culture and social acceptance rather than technical details. The ideals of beauty are very variable.

More Complex Modifications

Removal of Unused or Undesirable Genome

This is partially speculative, because currently we have very little understanding of the non-coding parts of the genome. Some parts (like promoters and various markers) are important for the function of the cell, while others are neutral or more or less destructive. Removed parts may need to be replaced with random or specially developed "fill out" DNA.

Transposomes

Transposomes make up around 10% of the human genome. While most of them are rather benign and have fulfilled an evolutionary function, they sometimes cause cancer and damage to vital genes. Since they do not have any biological function, the can probably be removed.

Oncogenes

There are several hereditary forms of cancer. One type is caused by defective anti-oncogenes, which prevent oncogenes from causing cancer. These can of course be corrected. Another type seems to be caused by already highly promoted oncogenes, which only require a slight push to become dangerous. These could perhaps be "tuned" to a more acceptable level.

Increase of Anti-Oxidant Enzyme Production

This could for example be done through promotion of superoxide-dismutase. This might help slow down aging and make the body more resistant to environmental dangers and free radicals. However, increasing the amount too much will probably interfere with normal biochemistry; further research is needed. One possible solution would be to add some ways of controlling the amounts, for example by linking them to the sleep cycle.

Improvements in Telomerase Activity

One of the more interesting theories about cell ageing is that the telomeres are gradually broken at each cell-division, until coding genome is destroyed after a certain number of divisions and the cell dies. If the activity of telomerase, which protects the telomeres, could be increased this would perhaps slow down cell-ageing.

One problem with this is however that it would increase the risk of cancer (many varieties of cancer have greatly heightened telomerase activity and are thus immortal). Improvements of telomerase activity must probably be combined with improved error correction and other anti-cancer enhancements to avoid increasing the risk of cancer too much.

Increase of DNA Error Correction

Some bacteria can survive very hard radiation, mainly through over-active error correcting enzymes. By increasing the amount and activity of similar enzymes (like DNA repair nucleases, AP endonuclease, DNA glyckosylases) in our genome, we would become much more safe from mutagens and radiation. In bacteria there are known enzyme complexes, known as the SOS response, which activate when the genome has been damaged. Unfortunately they increase the mutation rate, which is beneficial for bacterial survival but probably undesirable for multicellular animals, since they would increase the risk of cancer. Note that although evolution apparently can develop rather efficient protections against radiation and other mutagenic factors, it is normally not used more than at a rudimentary level in most living beings. There is no evolutionary advantage in being resistant against radiation in the low-energy environment on Earth, and the extra energy demands a lower fitness of most living beings. However, we humans have no problem increasing our energy intake as needed, and may have great use for better protection from radiation in space. Of course, no amount of error correction will remove mutations altogether, and it is probable that errors can be made in replication which are transmitted to the daughter cells.

Other Anti-aging Modifications

Beside the above mentioned possibilities (increased production of superoxide dismutase, error correction and telomerase), there are doubtless many other genes which could be optimized to improve the life span of the body. For example, certain forms of the APO-E protein seems to be linked with arteriosclerosis and Alzheimer's disease. If these are replaced with more efficient forms, this risk will be greatly reduced. It is probably hard to distinguish between removing damaged or disease-linked genes and prolonging the life span.

Production of New Substances

Genes could even be added for production of different substances (like vitamins, antibiotics or drugs), which could be activated by artificial hormones, special signals or chemical changes. In this case it might not even be necessary to modify the genome of all cells, just some suitable (like a patch of skin or the intestine) by a retroviral vector.

Resistance Against Poisons

It is also possible to add genes coding for enzymes breaking down or protecting against various poisons or irritants. Whether this is useful or not depends a bit on how paranoid one is about the chances of being poisoned. One application could be the production of chemicals binding environmental hazards such as heavy metals or carcinogenic substances. It might also decrease the risks of alcohol or drug abuse.

Complex Modifications

These modifications require quite extensive additions to the genome, deal with high level phenomena or require control systems.

Bacteriophage Genes

It might be desirable to attempt to add symbiotic bacteriophages to the genome. When activated by an artificial hormone or a bacterial toxin, the genes are expressed and the bacteriophages are produced. They will be used to seek out certain types of bacteria within the body or its cavities, and then attack them. This strategy might also include "tags" which makes the bacteriophages or the waste products of their attacks on intruders attract the attention of the immune system (the immune system will most probably limit the usability of phages to systems outside its reach, since it will regard the bacteriophages as an intruder). Care has to be taken to make sure the phage genes do not turn "rouge" within the genome or attack the cells unnecessary.

Genome Commenting and Marking

In order to improve the "legibility" of the genome, tags or markings could be placed in regular intervals or near important genes, so that they can be easily identified or changed. This will also reduce the risk for erroneous inserts.

Introduction of a Techno-Chromosome

Instead of placing new genes on the old chromosomes, it might be a good idea to introduce a new chromosome for this purpose only. This would decrease the risk that modifications cause undesirable changes to the rest of the system. The Human Genome Project already uses similar techniques to keep human genome libraries in yeast cells (so called YACs). The new chromosome would initially be filled with a noncoding pattern, with regular markers to simplify access or modification. One problem/possibility with adding another chromosome is procreation; unless the partner also has an extra chromosome fertilization will not work correctly. This will literally make the bearers of the chromosome a different species than Homo Sapiens. This could either be overcome by somehow designing the system so that the extra chromosome is not added to germ-line cells, or by using in vitro fertilization (which would probably be much more common, since the parents will most certainly want to determine what new genes to add and what to change). Some current research seems to imply that extra chromosomes can be turned on and off.

Additions to the senses

Several of our senses use chemical receptors to detect stimuli. These could probably be changed or expanded.

Sight - Currently the human eye uses three slightly different types of rhodopsin (also known as visual purple) for color vision. Other varieties are known among other animals, and could perhaps conceivably be added to expand the human perceptive range (in order to do this, the synthesis pathways must be added and placed near the other genes coding for rhodopsin synthesis and somehow linked to the cone-cell differentiation signals. Not exactly easy, but hardly impossible). This could expand the range of colors from the near ultraviolet (based on insect rhodopsin) to the near infrared. This would unfortunately work best on the embryonic level, since then the brain will naturally integrate the new type of cone to the visual system. Changes in adults would be much more unpredictable.

Smell - It is known that there are many chemical receptors used by the olfactory system, able to distinguish between several thousand types of compounds. It is probably rather easy to add new receptors, to recognize certain chemicals (like heavy metals) or perhaps other stimuli (long, unstable molecules could react to ionizing radiation, which would be felt as a certain smell). However, it is probably harder to train the human brain to handle smell than improving the sense organ itself, since we do not use our olfactory cortex to its full capacity, being very much visually and auditory oriented.

Taste - There are four known groups of taste receptors with several subgroups (It is interesting to notice their evolutionary value: salt signifies changes to the osmotic balance, sour the acidity balance, sweet is usually linked to high carbohydrate/energy content and bitter reacts to alkaloids/possible poison). Adding another group might present some differentiation problems like in the case of sight, and would probably be easier to do with additions to olfaction instead.

Artificial Symptoms

As Alexander Chislenko has pointed out in his essay about Enhanced Reality, many potentially life-threatening diseases lack easily noticed symptoms. These could perhaps be added, so that the patient will notice something is wrong on an early stage before it is too late.

A simple solution might be to add genes coding for enzymes producing a strongly colored compound, which colors the urine. These genes are normally repressed by a repressor which is inactivated by the presence of certain disease-indicative chemicals (several repressors could be linked, so that only certain highly selective combinations would cause the color-shift).

It would not be that hard to add a quite large number of such indicators to the genome, especially by using the techno-gene. Each could even code for a slightly dissimilar pigment, making diagnosis easier.

The most important use of this would be to detect cancer, which of course is a quite complex problem. One method would be to let the color-gene activate each time the cell divides; cancer cells would thus tend to have a different color (very useful for applying treatment) and tend to color the urine. The problem is of course designing the system so that normal cell division does not cause a false alarm. Another method would be to react to expression of known oncogenes, or perhaps known combinations of them (like loss of the expression of the epithelial cell-binding molecule E-cadherin and heightened expression of proteases, which signifies risk for metastasis).

Molecular Support for Cryonic Suspension

One of the main problems with current methods of cryonic suspension is the fact that ice crystals tend to disrupt the cells. However, certain animals contain anti-freeze proteins or fill the cytoplasm with carbohydrates which prevent the growth of large crystals. The genes for these systems could presumably be added to the human genome, decreasing the damage due to suspension.

Artificial Hormones

Many genes remain dormant until they are activated by the removal of a repressor protein or the binding of a promoter protein close to them. These genetic switches are sometimes controlled using signal substances or chemicals (such as lactose and glucose in the case of the lac gene), sometimes by more complex cascades of messenger proteins. It would not appear inconceivable that we could add similar systems to our own genes, giving us the ability to control which genes to express and which to inhibit.

By now you no doubt have an idea of what kinds of changes can be obtained simply by modifying the genome. We have not even touched on nanotechnology or synthetic biology yet.

Trent Goodbaudy

4 TRANSHUMANISM

Transhumanism is a meant to be used to refer to a transitionary period where a group of people want to eventually become the evolved and engineered species "Homo Evolutis" via the application of technology.

Transhumanism seeks to use radical advances in technology to augment the human body, mind, and ultimately the entire human experience as we know it. It is a philosophy that supports the idea that mankind will actively enhance itself, and steer the course of its own evolution.

Transhumanists wish to become what they call "post-human". A post-human is someone that has been modified with performance enhancing body and brain augmentations to the point where they can no longer be called human. They have mutated themselves into an altogether new being. To most people, this sounds like something from a science fiction novel. Few are aware of constant breakthroughs in technology which make the transhumanists mission a very real possibility for the near future.

Sold to us as beneficial, altruistic, with selfless arguments such as longevity. Who wouldn't want to extend their lifespan? Even if you are not transhumanist, many times you are friendly to the concepts of transhumanism. Advocates include those who wish to bring an end to suffering and disease. Maybe this research will help us to "turn off" cancerous tumors, or even some day provide a built-in protection to the organism against cancer. It sure sounds great to live a happier, healthier life where we are stronger, and perhaps even have superhuman abilities compared to an original Homo Sapien.

Transhumanism is a relatively new and exotic idea, and it has even become a cultural movement, it can also be thought of as the ultra high-tech dream of scientists, philosophers, neuroscientists as well as many others. This concept gained significant popularity when the mapping of the human genome was achieved, now already more than two decades ago. It has to do with the concept of transgenics, (***Transgenics*** - 1. Of, relating to, or being an organism whose genome has been altered by the transfer of a gene or genes from another species or breed: transgenic mice; transgenic plants. 2. Of or relating to the study of transgenic organisms: transgenic research.) and crossing over of species barriers, changing our bodies via genetic modification, nanotechnology, artificial intelligence, and synthetic biology among other technologies.

Transgenics research is currently being funded by the United States Department of Health, and the Defense Advanced Research Projects Agency (DARPA) to the tune of millions of dollars currently in publicly allocated and congressionally approved funds. DARPA already has ongoing projects in the; bio-design, nanotechnology, synthetic biology, and artificial intelligence categories. The Air Force, Navy, and Army are also currently funding this research. Other governments around the world are also funding research in these exotic fields; including China, Australia, the United Kingdom, and others.

Likely billions or even trillions of dollars have been and are currently allocated to other non-publicized government "black" projects, as well as private/corporate research on this matter. It is frightening just to think of some of the results of some of this "research". Although the scariest prospect might be to imagine what kind of things are happening in the "rogue" labs that the government is concerned about competing with, and staying ahead of. The prospect of an entity using this technology as a weapon of mass destruction is not out of the scope of reality either.

As you may have already guessed, the military is very interested in these recent technologies for development of exotic new materials and ultimately an immortal, super-soldier, war-fighter, machine hybrid that is neither completely human nor animal nor machine, but has the best qualities of each for fighting. As a matter of fact, all branches of the military are actively researching and advancing this technology. The way they look at it, there is private research going on in this field all over the world right now, and to fall behind (in their view) is to give the advantage to the enemy. It is vital to national security, and perhaps even the very survival of mankind, to be the leaders in this field of research. I am afraid that this is a very strong motivation to continue this research by both the military and it's adversaries, and in my opinion we are possibly approaching a major catastrophe, and maybe even an extinction level event unless something can be done to control the advancement of these fields.

As an example, we presumably already know that a spider's web is one of the strongest crystalline fibers produced in nature, and it is old news now that transgenics was used to engineer a goat so that silk could be harvested from the milk of the transgenic spider-goat. The transgenic goat was altered to produce a form of spider web in the milk when it lactates, for use in manufacture of body armor, parachutes, cargo nets, and in many other applications.

> *"Researchers from the University of Wyoming have developed a way to incorporate spiders' silk-spinning genes into goats, allowing the researchers to harvest the silk protein from the goats' milk for a variety of applications. For instance, due to its strength and elasticity, spider silk fiber could have several medical uses, such as for making artificial ligaments and tendons, for eye sutures, and for jaw repair. The silk could also have applications in bulletproof vests and improved car airbags."*
> - PhysOrg.com (May 2010)

This type of technology is even more sophisticated now, and continues to advance at an exponential rate. Scientists admit that they are now "comfortable" creating humanoid organisms with 50% non-human DNA. These scientists have already made clones, genetically modified pretty much everything, and now they just need to gradually introduce us to, and get us as acclimated and comfortable with their "creations" as they get older and grow into maturity.

Pharmaceutical companies are also very interested in a better alternative to the mind altering drugs of today. They often have to labor with the Food and Drug Administration over testing of new drugs and technology. So what they do is create hybridized chimeras via transgenics and perform research on animals with implanted human DNA.

We may have had problems previously transplanting organs from animals into humans, but scientists can now just grow organs inside a part-human transgenic pig for example. These organs are much less likely to be rejected, because they are not entirely made of animal DNA, they are made of human DNA grown in a chimera. The same theory goes with testing genetic "medications"; if you can test a medication on an animal with human DNA then you don't need human subjects anymore for testing. On a sarcastic note; I am sure that the pharmaceutical industry would be eager to sell us a replacement to subversive experience with genetic gradients of pleasure that keep humanity in a state of constant bliss, of course with an ever improving method of "non-invasive" delivery.

Chimeras are also created for stem-cell research, there is no longer any need to get involved with the controversial topic of obtaining stem-cells via fetal tissue. As a matter of fact, there is even research going on in the area of human reproduction as well that completely omits the inconvenience of pregnancy and childbirth. Originally the transhumanists had an idea for childbirth that consisted of a clone that is immediately anesthetized so that it never knew it was alive. This clone would be used for the duration of the pregnancy. You may see the moral and ethical issues that could arise with this method, so it was revised so that all that would be needed to be cloned would be the womb. This cloned womb could be integrated into a synthetic AI device that would have a heart-beat, play audio recordings (recorded by the surrogate "mother"), and this robotic womb could also move around simulating a real womb environment, giving the growing genetically engineered baby a realistic and nurturing nine months in the womb.

Some foresee an Earth in which natural reproduction is viewed as irresponsible, with newer methods of reproduction emerging constantly, soon it will be the "responsible choice" for both society (since you're selecting only the best genes) and your children (since you're choosing what is best for them). Others, such as author John Glad, view eugenics as "an integral component of an environmentalist policy" (Future Human Evolution, 2006). He continues, "Abortion should be actively promoted, since it often serves as the last and even only resort for many low-IQ mothers."

President George W. Bush signed legislation to limit funding of cloning and transgenic human research as early as 2001.

In the 2006 State of the Union Address, President Bush said *"A hopeful society has institutions of science and medicine that do not cut ethical corners and that recognize the matchless value of every life. Tonight I ask you to pass legislation to prohibit the most egregious abuses of medical research: human cloning in all its forms; creating or implanting embryos for experiments; creating human-animal hybrids; and buying, selling, or patenting human embryos. Human life is a gift from our Creator, and that gift should never be discarded, devalued, or put up for sale."* (American Presidency Project: http://www.presidency.ucsb.edu/ws/index.php?pid=65090#ixzz1mvK1aqOs)

Yet all this effort only stalled the process for a few years, as when President Obama obtained office he overturned Bush's restrictions on experiments with stem-cell research and chimeras.

"President Barack Obama ended limits on funding for embryonic stem-cell research set by President George W. Bush in 2001. The announcement Monday was coupled with the signing of a presidential memorandum directing that "scientific integrity" be restored to government decision-making.

"Rather than furthering discovery, our government has forced what I believe is a false choice between sound science and moral values," Mr. Obama said. "In this case, I believe the two are not inconsistent. As a person of faith, I believe we are called to care for each other and work to ease human suffering. I believe we have been given the capacity and will to pursue this research and the humanity and conscience to do so responsibly." - Wall Street Journal (March 2009)

Human Rights for Non-Humans

Discussions about government funding and research of genetic engineering of the human genome will eventually force another topic into the public realm. How should the Bill of Rights be extended to "Human Non-Humans"? What percentage of human DNA does an organism need to contain to be considered human? If someone (or a transgenic human organism) was not "legally" defined as human, could they then be discarded or treated differently than a "wild" or natural human?

Working on a project funded by the United States Department of Health through Arizona State University; Professor Maxwell Mehlman has two presentations available currently on the Arizona State University Website on this very subject.

They are entitled:

"Improved Humans: Legal and Political Aspects of the New Genetics"
Templeton Research Lecture
November 12, 2008

and

"Directed Evolution: Public Policy and Human Enhancement"
Templeton Research Lecture
April 20, 2009

In other words, this research is really already being funded, and is going on behind-the-scenes, without discussion in the public light. I would think that this subject matter should be brought to light, before it is too late.

> *"The moral challenge of transhumanism will transcend those of abortion and euthanasia. For this reason, the pro-life movement must become the pro-human movement."* —Nigel M. Cameron

Nigel M. Cameron sees genetic manipulation as leading to a *"new feudalism,"* wherein a *"very small number of people, basically a global elite,"* will take advantage of a *"law of compounding,"* using their genetic advantages to create a society with *"far greater disparities"* in wealth and power than currently exist. Once a certain proportion of the population has had fundamental genetic or mechanical enhancements, these societal changes will become, he says, *"absolutely inevitable."*

What stance would religion take on these matters?

Do transgenic humans have a soul?

One might wonder why an individual would be interested in an "after-life" when they could extend their current life for as long as they wanted. There are, however many religious transhuman associations; including a

Christian Transhumanist Association, a Buddhist Transhumanist Association, a Muslim Transhumanist Association, a Mormon Transhumanist Association, even a Non-dogmatic Transhumanist Association, among others.

Several notable and opinion-leading transhumanists are strongly anti-Christian however. For example, prominent transhumanist William Sims Bainbridge, the author of more than 15 books and numerous magazine and journal articles, opens an article with the following abstract: "*Cognitive science immediately threatens religious faith in two ways, by explaining away religion as an error resulting from accidents in the evolutionary history of the human nervous system, and by failing to find evidence that humans possess souls. Over the coming decades, information technology may undercut people's need for religion by offering practical forms of cyber immortality (CI). The plausibility of religion may also be eroded by the coming unification of science and the associated convergence of* [new] *technologies.*"

Even more troubling is language on the Web site for the Future Technologies Advisory Group (www.futuretag.net), a transhumanist organization specializing in consulting and media: "*While one of the objectives of the firm will be facilitating the penetration of transhumanist ideas in mainstream business and policy, we will not use the T word or insist on the transhumanist worldview too explicitly. Rather, we will focus on delivering practical advice appropriate to the intended audience.*" The willingness expressed here to dissemble their true intentions is disconcerting.

And yet there must be open and honest dialogue on these issues. "*We are moving way beyond these old challenges to human life,*" argues Cameron, who says we need to change our focus of attention. We need to "*ask the right questions*" and work toward finding ethical answers—answers consistent with *some* moral tradition—before the future arrives.

The Vatican has also been emphatic in its stance against nearly all eugenic plans and techniques. Even as far back as 1987, the Congregation for the Doctrine of the Faith's "Instruction on Respect for Human Life" stated, "*Certain attempts to influence chromosomic or genetic inheritance are not therapeutic but are aimed at producing human beings selected according to sex or other predetermined qualities. These manipulations are contrary to the personal dignity of the human being and his or her integrity and identity. Changing the genetic identity of man through the production of an infrahuman [i.e., inferior] being is radically immoral. The use of genetic modification to yield a superhuman or being with essentially new spiritual faculties is unthinkable.*" More recently, this teaching was rearticulated by Castrillón Cardinal Hoyos late in 2006, when he said, "*Genetic manipulation, when it is not therapeutic, that is, when it does not tend to the treatment of pathology of the genetic patrimony, must be radically condemned. . . . It pursues modifications in an arbitrary way, inducing to the formation of human individuals with different genetic patrimonies established according to one's discretion. Eugenics, the creation of a superior human race, is an aberrant application.*"

Still, there are some gray areas—for instance, in the definitions of "*therapeutic*", "*reparative*", and "*augmentative*" gene therapy. Award-winning scientist and author Ray Kurzweil — an avid transhumanist and proponent of radical life-extending technology — when asked if there was a difference between genetic therapy for a person with Down Syndrome and for a person who wanted an IQ of 135 instead of 100, responded, "*In my opinion, no. We are the species that goes beyond our limitations.*" One could take a similar approach along theological lines, but with a view toward licitly undoing the effects of the Fall through technology.

New Modes of Perception

This road down transhumanism will introduce new possible modes of being, and new human perception. On his website (nickbostrom.com); Nick Bostrom states the following;

> " *The range of thoughts, feelings, experiences, and activities accessible to human organisms presumably constitute only a tiny part of what is possible. There is no reason to think that the human mode of being is any more free of limitations imposed by our biological nature than are those of other animals. In much the same way as Chimpanzees lack the cognitive wherewithal to understand what it is like to be human – the ambitions we humans have, our philosophies, the complexities of human society, or the subtleties of our relationships with one another, so we humans may lack the capacity to form a realistic intuitive understanding of what it would be like to be a radically enhanced human (a "posthuman") and of the thoughts, concerns, aspirations, and social relations that such humans may have.*

Our own current mode of being, therefore, spans but a minute subspace of what is possible or permitted by the physical constraints of the universe (see Figure 1). It is not farfetched to suppose that there are parts of this larger space that represent extremely valuable ways of living, relating, feeling, and thinking.

The limitations of the human mode of being are so pervasive and familiar that we often fail to notice them, and to question them requires manifesting an almost childlike naiveté. Let us consider some of the more basic ones." - Nick Bostrom - Professor, Faculty of Philosophy & Oxford Martin School - Director, Future of Humanity Institute

The Space of Possible Modes of Being

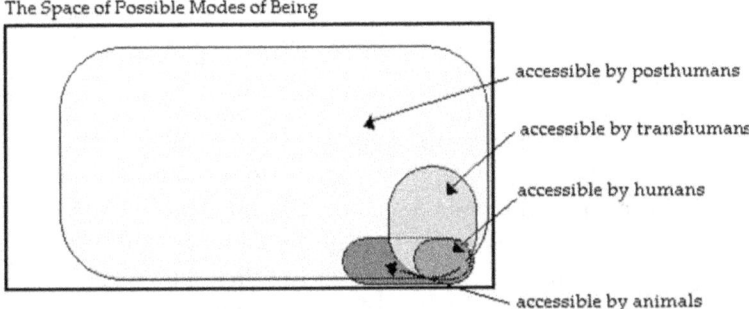

Figure 1. We aint seen nothin' yet (not drawn to scale). The term "transhuman" denotes transitional beings, or moderately enhanced humans, whose capacities would be somewhere between those of unaugmented humans and full-blown posthumans. (A transhumanist, by contrast, is simply somebody who accepts transhumanism.)

In the above figure, Nick does an adequate job symbolizing the vast modes of being that would be possible to "posthumans". This really looks like we really will have the opportunity to become superhuman; or at least we will have super-sensory perception.

Imagine being able to "smell" a tumor. How would we experience the world with superhuman senses? Have you ever heard stories where animals were able to see something and perhaps even react to something that we could not detect? A dog that is usually fearless, running scared and crying of something unseen to us. I am sure that I am not the first to wonder if animals can see into the spiritual realm.

What if we, as posthumans could see new spectrums of light (infra-red and ultra-violet) along with super enhanced sight? Would we also perhaps be able to cross over or at least see into the ethereal realm? This path could also lead to other "supernatural" abilities such as the ability to communicate with alternate dimensions or even the ability to fully realize our own existence and learn to teleport, or travel through time, or dimensions. Would we be able to make contact and communicate with the dead, angels, aliens, or demons?

For a very long time cats and owls have been associated with witchcraft, used as mediums, and it is rumored that they can contact or at least "see" through to the ethereal realm. What if we gave ourselves the best that our animal kingdom has to offer, and on top of that gave ourselves immortality, and then combined that with the best of man-made nanotechnology and artificial intelligence? I venture that we would essentially have become gods. But we still do not know if the Human 2.0 will even have a soul.

Bio-Chip Implants

Human-machine interfaces are becoming ever more intimate. Amazing progress has been made in integrating technology with biology, progress that has helped people tremendously. For example, former football player Jesse Sullivan, who lost his arms in a utility line accident, now has two bionic arms that can move in response to his thoughts. Claudia Mitchell, who lost an arm in a motorcycle accident, now has a prosthesis so precise that she can peel an orange. Researcher William Craelius, working on a different track, has developed the Dextra, an artificial hand that allows users to type and even play the piano. The next generation of limbs will even allow wearers to sense touch and temperature.

Going a step further, consider the case of quadriplegic Matthew Nagle, who "*can now pick up objects, open e-mails, change the channel on the television and play computer games*" using only a link between a computer, a robotic arm, and electrodes implanted in his brain, according to London's Independent. Scientists are also hard at work on a wearable exoskeleton that would respond to his thought commands, allowing him to move his body again.

But what if technology is doing more than simply correcting a medical condition? Dr. Steve Mann of the University of Toronto has been called "the first cyborg" by DK Publishing for his "WearComp"—wearable hardware that runs personal-applications software. "*The assumption of wearable computing*," reported Mann at a 1998 keynote speech, "*is that the user will be doing something else at the same time as . . . the computing. Thus the computer should serve to augment the intellect, or augment the senses."*

Ray Kurzweil has pointed out that, when it comes to computer interfaces, "We already have people with computers in their brains—for example, Parkinson's patients—and the latest generation of this FDA-approved neural implant allows you to download new software to your neural implant from outside the patient. . . . In the future we will have non-invasive ways of extending our physical capabilities as we merge with nonbiological systems." Interesting that he says this, I do wonder what he means by "non-invasive". Could he be talking about synthetic biology or trans-biology?

So where do we draw the line? At what point does the interface between human and computer present a challenge for human personhood? The ultimate goal of a brain chip would be to increase intelligence thousands of times over. Which could theoretically turn the brain into a super computer. Imagine learning a new language or skill instantaneously.

Lifelong emotional well-being is a fundamental concept among transhumanists, this can be achieved via re-calibration of the pleasure centers of the brain.

Nanotechnology is a pivotal area concerning transhumanists; it is the science of creating machines that are the size of molecules. Nanotech machines will be able to build at an atomic level, and replicate ANY chemically permitted form of matter. Such machines could create organic tissue for medical use. Using this type of technology could dramatically improve lifespan. Some experts say that it will soon be possible to live... *forever.* Artificial Intelligence, or the creation of thinking robots is also closely related to the mind-machine merger concept of neurochips.

Much speculation is being done as to what relationships will exist between advanced humans and AI machines. Some believe that machines will completely take over, and others believe that machine parts will be added to our bodies to create cyborgs, and still others believe that humanity will merge totally with technology, by uploading individual consciousness to a virtual reality. Upon being downloaded, one could live forever in a computer generated reality, leaving the physical body behind. In this machine, the individual can merge their intelligence with the collective intelligence of all others in the digital reality effectively becoming one super-intelligent being.

This concept is popularly referred to as the "hive mind". These scientific breakthroughs sound so amazing and advanced that it makes one wonder if they can actually be done. But I believe a more important question is; should they be done? It is understandable that people want to have happy, healthy, abundant lives with as many years to experience as possible. The concept of a better world for the common person is what fuels the transhumanist movement. However, surveying the world of today without thinking of dreams for tomorrow reveals some cold hard realities. The main one being that the world as we know it is run by a small sociopathic group of elites who are anything but generous. This has been the case throughout history; the idea that such technology would be shared with common people contradicts the lessons of the past.

If the technology were to be given to common people, an even greater danger exists. Imagine a real life "wolf-man" on the loose, killing and raping victims. Or even real life "super-villains" that at one time only existed in comic books.

The science of controlling populations can be fully realized through the use of brain chips and through hybridization of our DNA. This is a possible control freak's dream come true, a completely digitized population that could be manipulated in the same way one controls a computer. Just who will be behind the post-human keyboard?

The transhumanist agenda caters to a materialistic mind, it offers us the key to emotional well being, superior intelligence, and prolonged existence that completely ignores the fact that these things are readily available to us right now.

Science has tried to drum out our build and spiritual connection to the universe and replace it with a lacking version of objective reality in which the facts speak for themselves. Our rational minds are anything but rational because they automatically assume that something is one way or another. We can either be religious, or scientifically minded, there is no third alternative. You are either this or that, one or the other. It is through this lack of personal investigation that we lose sight of ourselves and decide to follow established lines of thought.

We ally ourselves with movements, rather than empower each other's unique individuality, and thus the hive mind is born. Do you really want to live forever within a machine? And who is to say that we aren't already in such a machine searching for a way out? What we need is a true understanding of spirituality, not a high and mighty techno-future solution to our perceived worldly problems.

Problems with transhumanism

With benefits such as longer life, increased intelligence, cognitive enhancement, increased performance, and better communication it's easy to see the appeal of transhumanism. But along with these benefits come the potential dangerous applications of converging technologies. Dangers which are either overlooked or completely ignored by the transhuman fundamentalist community. At an intuitive level I think most people are well aware of these dangers however they are not aware of transhumanism in general.

The following list of problems will only increase in relevance as transhumanism steadily makes its way into the mainstream.

Too positive - The way that this is being sold to us, and incrementally being introduced (almost in a sinister way); it sounds way too positive, too utopian. What is that old saying? Oh, I remember... *"If it sounds too good to be true ... it probably is."* You know that we already LOVE our digital toys. When this stuff starts making an appearance in the health/beauty/retail marketplace, it is human nature to exploit, take advantage of, and even abuse our technology. I am afraid that we would not be able to control ourselves, and I have a feeling that the military is not going to be able to satiate their addiction to this technology either.

Genetically modified food - By crossing over genetic barriers we are already feeling some of the unintended consequences. What if human DNA infects the food supply? Would we find ourselves afflicted with a human form of mad cow disease? You remember that cows got the "mad cow" condition by consuming their own DNA, right? Maybe this was another little detail that the media left out of their story. There have also been reports of people having respiratory conditions that live downwind of GM crop fields.

"There are some research studies that have linked genetically modified foods with allergic lung damage in mice and the creation of antibodies specific to a food protein. At present, the US Food & Drug Administration (FDA) doesn't mandate pre-market safety tests. However, it does review companies' voluntary GE/GMO toxicology, nutrition, and allergenicity testing results. So a soybean can't be genetically modified to include the protein of a peanut, yet there are many ways the soybean is currently being modified. A gene was added to selected soybean varieties that make them virtually immune to damage by Roundup herbicide, which will kill most other plants. Of course, we the consumers are then eating this modified soybean."

On January 28, 2007, The New York Times Magazine's lead article *"The Age of Nutritionism"* discussed *"how scientists have ruined the way we eat."* Their suggestion was to *"eat what our great-great-grandparents would have recognized as food! Scientists have found ways to create waffles with Omega-3's in them, however no one knows whether you're really getting the quality of Omega-3's that you would get by just eating fish. Chances are something got altered in the science. Eating the original food provides an entire system of nutrients that may need to act and react together for you to get the advantage of the nutrient. Isolating a protein or a nutrient probably doesn't give us the same end result.*

The Grocery Manufacturers of America estimates that between 70 percent and 75 percent of all processed foods available in U.S. grocery stores may contain ingredients from genetically engineered plants. Breads, cereal, frozen pizzas, hot dogs and soda are just a few of them. If your family is eating any processed foods, it is highly likely that you are eating genetically modified foods. The most common genetically engineered crops in the US are corn, soybeans, cotton and canola. There is also a genetically engineered hormone, BGH, which is commonly injected into dairy cows in the United States." -
http://www.allergicchild.com/causes_food_allergy.html

"The FDA has no information that the use of biotechnology creates a class of food that is different in quality, safety or any other attribute from food developed using conventional breeding techniques," says James Maryanski, Ph.D., the FDA's food biotechnology coordinator. He adds that disclosure of genetic engineering techniques is not required on the label, just as identification of conventional breeding techniques is not required-- for example, "hybrid corn" can just be called "corn."

> *"On August 10th, 1998, Arpad Pusztai of the Rowett Research Institute in Aberdeen, Scotland appeared on the British TV show "World in Action."* In the course of the interview, he announced that his experiments showed that rats fed a diet of potatoes expressing a gene coding for a snowdrop sugar-binding protein showed stunted growth and reduced immune function (Enserink, Science 281.1184). He is further quoted as saying that he would not eat GM food and that he found it *"very, very unfair to use our fellow citizens as guinea pigs"* (Lee and Tyler, 1999). - http://hils.psu.edu/lsc/fedoroff.html

Rats will not eat GM potatoes unless they are absolutely starving, what do they know that we do not?

> *"When food-crops are genetically modified, ("genetically modified" food is a misnomer!) one or more genes are incorporated into the crop's genome using a vector containing several other genes, including as a minimum, viral promoters, transcription terminators, antibiotic resistance marker genes and reporter genes. Data on the safety of these are scarce even though they can affect the safety of the GM crop."* - http://www.actionbioscience.org/biotech/pusztai.html

We are only just beginning to feel this issue currently, and it is only going to get worse unless we speak up about it. Speak with your dollars, that is what they listen to best. Start growing your own food or obtain it from local reputable farmers. The term "organic" is being obfuscated, Whole Foods is being forced to carry genetically modified foods and calling them "organic", so it is best to find an alternative to the grocery store for things you are going to consume. If you do not have a local farmers market, the least you can do is try and get the ball rolling on this and start making some calls and try to organize something. We have to help each other. We need to stand up and be heard.

Life extension - the desire to lead a long happy healthy life is undeniably a good thing, but if we introduce the ability to extend human life indefinitely certain issues will arise. Already we have been inundated with claims that the earth is overpopulated. The correlation between environmental degradation and rising population numbers is not new.

> "The natural world in which man lives and in which we depend is indeed deteriorating... is being destroyed in many instances at a rate that is accelerating and can only continue to accelerate unless we begin to control the activity that was used to start the impact - It took mankind something like 50 millennia to achieve its first billion of population on this earth, now it takes us some 20 years to add a billion people. Now this has got to be regarded as a serious problem. This might be necessary at some point... at least some restriction on the right to have a child, I am not proposing this I was simply predicting this as a... as one of the possible courses that society would have to seriously consider should we get ourselves into this kind of situation."
> - Maurice Strong 1972.

China phased in it's one child policy gradually, and it makes sense that similar policies would be phased in globally as life extension therapy has become widely used. In a world where humans live forever, who gets the final say on reproduction? Government would manage human life. The right to life could become a privilege based to some degree on each person's degree of usefulness to society. This scenario is unavoidable if radical life extension was widely employed. This brings up deep philosophical issues which cannot be ignored.

Law Enforcement & Crime Scene Analysis - Crime scene investigators would have some catching up to do. As a matter of fact, they would have to take an entirely new approach to crime solving. Imagine that real life part human "wolf-man" I mentioned earlier. The assailant is essentially an animal, they would have a different mindset and instincts. Profiling and forensics would either not apply or be totally different. And once it was tracked down; how would law enforcement bring them to justice? Would they even have any human rights? What about future terrorism by transhumans? Or the opposite; what if we ended up with an infestation of "super-animals"? What kind of a new arms race would exist with this technology? The government is being FORCED down this road; it is not a matter of ethics, it is a matter of winning.

Implantable chips - implantable brain chips are feasible. The nature of their future implementation has already been discussed in government documents. Although brain chips could enhance mental capacities dramatically, they could also be used as a means of outside control. Thoughts, sensory stimulations and false memories could be instantly "beamed" into your mind and create a totally synthetic reality which would be perceived as real. Humans could be remotely controlled and manipulated. Denying that this would happen does not mute it's possibility.

Mind uploads - uploading your mind to a computer brings an alternate way of achieving immortality, but it presents limitless possibilities for malevolent use. Leaving your material body behind to become an electronic data string could quickly bring limitless digital experiences, the problem here is perception. Yes, you could feel powerful, but the reality would be that your entire existence is contained within a tiny computer. Computers are programmed. Whoever programmed the computer would have complete power over your life. If they wanted to, they could manipulate and torture you in ways that we cannot even conceive. It would be beyond your control to stop them, because you would not be able to return to the material world. You would be trapped. The irony of this situation is that you are leaving your body in pursuit of a pseudo-spiritual existence you give the power of godhood over to those people you leave behind in the physical world.

Artificial Intelligence - the problem with artificial intelligence is its potential to become infinitely more intelligent than us. By design, an AI brain would not be human. Would artificial consciousness recognize and respect it's creator? And after watching Arnold Schwarzenegger in The Terminator, it really doesn't make sense that its ultimate goals would parallel those of mankind. It seems impossible to keep AI under human control, just because machines would be more intelligent than us, does not mean that they will be superior to us. If we choose to create artificial intelligence, we may be opening the gate to a realm of imperceptible consequences.

Extinction - There is no denying that we could go too far and end up doing something really bad to humanity. For example; this was the premise of the movie "I am Legend" with Will Smith. In the movie, a cure for cancer was genetically engineered and the vaccination was mandated to all of humanity; however the scientists did not count on the entire world coming down with a genetically modified "super-rabies" that was a result of combining animal and human genes, swept through the vaccinated masses.

Of course, none of these technologies will arrive in their advanced forms immediately, they will be introduced over time so as not to completely overwhelm everyone. This in itself is a major problem, because as we acclimate to them, we could easily take them for granted. The dangers won't go away, in fact they will only increase in relevance. This illustrates one of the biggest problem with transhumanism in general; it focuses on future miracles and ignores present realities. The truth of the matter is that our world has major problems, problems that can't be cured with external remedies. We cannot build a utopian society upon a broken foundation. If we choose to march forward without first looking inward then we will surely bring about our own undoing. We are riding the bike while we are building it! We are creating a molecular, biological nightmare. What will the impact be on humanity, and even the world? You will only hear about little bits and pieces of information from the mainstream media; this subject is essentially censored from the public view. The public does not need to be concerned about this anyway, trust your government and your military and your pharmaceutical industry, they know what is best for us. Right?

We are throwing off the balance of nature. There is a divine order for a reason, I feel that there are also barriers to the earth's species for a reason. We should not play with these forces, this is very serious. Even evolution does not talk about this, we should proceed with the most extreme caution, if we must proceed at all.

Darwin never even dreamed of this.

What does the future hold for humanity?

5 NANOTECHNOLOGY

At Lawrence Berkeley National Labs and now all over the world, researchers are manipulating particles at the atomic level. This technology is ushering in new potential cures for cancer, clothes that don't stain, and solar panels as thick as a sheet of paper.

The common unit of measurement in nanotechnology is nanometers. One nanometer is equal to one billionth of a meter. One way to put this into perspective is to think of a human hair; a nanometer is still 100,000 times smaller than the width of single strand of hair. This is very small indeed. The ability to observe and construct things this small is at the heart of nanotechnology. Scientists have discovered that at the nanometer scale; everyday materials start to act in unimaginable ways.

> *"The behavior of nanomaterials changes or can change when the size becomes so small. When compared with a larger amount of that same material. When you have things that sort of start changing the way they behave, and now you have the ability to control that, it sort of opens up an entirely new base of material"*
> - Jeffrey Grossman, Executive Director of Integrated Nanomechanical Systems, UC Berkeley

The fact that you can customize nanomaterials unique behaviors has already turned "nano" into the buzz word of the decade. Some researchers predict nanotechnology could lead to faster computer chips, tiny medical devices, even new filters to clean water pollution.

As novel as nano materials seem, they have already been in use by humans for hundreds of years. For centuries, the colors in stained glass windows for example have been the result of a controlled heating and cooling process that adjusts the size of tiny crystals in the glass. Stained glass is essentially medieval nanotechnology. What is different now though is that we have the ability to see what we are doing at the nano scale, and what is happening. This offers an ability to design materials rather than to just kind of find them accidentally. Grossman goes on to say: "*But understanding and controlling how nanomaterials act can be tricky. If you take a piece of a material, say silicon, and you look at that material, it looks kind of dark charcoal and is not very interesting to look at. Now if you take out you little nano "ice cream scoop" and you scoop out a nano-sized chunk of that material and you look at that; all of the sudden it glows blue, and if you take a slightly bigger ice cream scooper out ... it glows red. So now you have a material that completely changes the way it looks, the color that it is, just by changing its size.*"

So then the question is why do nano-materials behave so strangely? A simple explanation is that when you start working with materials at this size; in the case of silicon, the electrons essentially do not have enough room and they move faster creating a glow. It is known as "quantum confinement". The smaller you make the crystal, the higher the energy of the electron will be. It's kinetic energy has increased, and that can also be thought of as making its "wavelength" a bit shorter, and forcing it into a box where it kind of zips around more quickly.

The relationship between volume and surface area is another factor that shapes nano-scale behavior. Things this small have much more "outside" than they have "inside". The surface area of the material starts to skyrocket compared to it's volume, and in fact when you get down this small; most of the material could be just surface, and very little of it is actually volume. The reason why this is interesting, is because the more surface you have, the more reactions you can carry out on that surface. Which allows you to do things such as filtering water more effectively, for example.

Our ability to observe and change things at the nano-scale has led to a host of new materials applications, and has even made it possible for scientists to build nano-mechanical machines. Like this nano-motor pictured below.

From memx.com: "*Imagine a machine so small that it is imperceptible to the human eye. Imagine working machines no bigger than a grain of pollen. Imagine thousands of these machines batch fabricated on a single piece of silicon, for just a few pennies each. Imagine a world where gravity and inertia are no longer important, but atomic forces and surface science dominate. Imagine a silicon chip with thousands of microscopic mirrors working in unison, enabling the all optical network and removing the bottlenecks from the global telecommunications infrastructure. You are now entering the microdomain, a world occupied by an explosive technology known as MEMS. A world of challenge and opportunity, where traditional engineering concepts are turned upside down, and the realm of the "possible" is totally redefined. MEMS are quietly changing the way you live,*

in ways that you might never imagine. The device that senses your car has been in an accident, and fires the airbag is a MEMS device. Most new cars have over a dozen MEMS devices, making your car safer, more energy efficient, and more environmentally friendly. MEMS are finding their way into a variety of medical devices, and everyday consumer products."

Even though scientists can now build working machines at the nano-scale, nature is still king.

"Some of the most fascinating things that we see in the nano-scale are the machines that nature builds at the nano-scale. Nature makes rotors and motors and other things that operate on this scale of a few nanometers by making protein assemblies and that have interesting nano-mechanical properties. If you look at the flagellum of a bacterium it is a little nano-scale motor, and it has bushings and all kinds of little pieces inside there. They are just built differently from the way we make ours. And they have some tricks that we are trying to decipher." - Paul Alivisatos, Director, Material Science Division, Lawrence Berkeley National Lab.

Alivisatos and his colleagues are eager understand and apply some of these new ideas to the real world. Among their goals is finding new forms of clean energy. Every minute, enough of the suns light reaches the earth to meet the world's energy demand for an entire year. This is a logically good place to start.

"Photosynthesis and photovoltaics is surely the one [area] *where we have the most interesting conversation that goes on, between how nature achieves this, and how might we do it in the future. If were to get our energy from the sun, we need to copy some of the lessons of nature, but we also need to find some of our own little tricks that we can apply."* - Paul Alivisatos

Along with promise of a bright new future, nanotechnology brings with it, it's own share of risks and fears. In December 2006, the city of Berkeley amended its hazardous material law to include "nanoparticles". Making it the only local government in America to regulate nanotechnology.

"*We know that nano-scale materials can enter inside cells, and we know that that could have consequences to health. So, it's incumbent, it's really required that we do research to understand; what is the nature of the interaction between new engineered nano-scale materials and living systems, not just cells, but whole living beings.*", said Alivisatos

~

Meanwhile; at University of Nebraska nanotechnology is taking on a whole new meaning with the hybridization of nanomaterials into biological material such as DNA and RNA:

"Scientists are always seeking better ways to find and quantify minute things, such as toxins in the air or cancer particles in blood. UNL researchers lead a collaboration to create more powerful detection devices by combining manmade nanoparticles with nature's inherent recognition capabilities.

Creating these "nanohybrids" requires the diverse expertise of researchers in biology, chemistry and nanomaterials engineering. A Nebraska team recently launched the UNL-based Center for Nanohybrid Functional Materials, which brings together 15 researchers from UNL, the University of Nebraska Medical Center, the University of Nebraska at Kearney, Creighton University and Doane College.

With nanohybrids, "you get the best of both worlds," said UNL chemist Patrick Dussault, a Charles Bessey Professor, who co-leads the center with Mathias Schubert, associate professor of electrical engineering.

Nanohybrids combine nanostructures – which can be engineered to behave uniquely under certain conditions, such as when subjected to a beam of light or radio energy – with chemical or biochemical agents, such as RNA or antibodies that can bind a specific substance. This new nanomaterial can both find and reveal its target.

Materials often behave differently at nanoscales, Dussault said. Understanding the basic sensing principles of nanohybrids is a

major goal of the new group. With this knowledge, researchers hope to develop tools with enhanced detection capabilities.

Potential applications include devices that more selectively or sensitively diagnose diseases or find environmental contaminants. The ability to better detect toxins in air or water also could benefit national security.

The center builds on UNL's strength in nanomaterials. With about $7.5 million in funding from the National Science Foundation through Nebraska EPSCoR, the center is creating a new core facility and partnering with several departments to hire new faculty, enhancing UNL's leadership in nanoscience.

The center also has begun developing partnerships with industries in Nebraska and beyond.

"I think potentially it can attract a lot of companies, big and small, to Nebraska," said Fred Choobineh, Nebraska EPSCoR director. "It's very creative and cutting-edge research." - Nanohybrids Promise 'Best of Both Worlds' - Annual Reports 2010-2011 from the Office of Research & Economic Development - University of Nebraska–Lincoln

Self-replicating nanostructures

When the impact of nanotechnology is fully realized, amazing things will be possible.

DNA is at the leading edge of nanotechnology and is the basis of the construction of many two and three-dimensional nanostructures. Now DNA nanotechnology researchers are able to fold DNA molecules into a wide variety of structures to create nanoscale DNA tiles. These nanotiles are building blocks for nanoarchitectures to design and fabricate a broad range of structures and devices. DNA is an ideal macromolecule to be utilized in the quest to develop new self-assembling nanostructures. Molecular self-assembly is a novel strategy of choice for developing new nanostructures; it is a spontaneous, noncovalent association of specific molecular aggregates, and is a consequence of the molecules' inherent physicochemical properties. DNA has now itself turned into a molecular construction material with a great number of applications beyond genetics and nanotechnology. Capable of virtually limitless applications, ranging from sensitive molecular detection to DNA-based computing, and fabrication of nanoelectronic components.

In 1992, seminal nanotechnology pioneer Dr. K. Eric Drexler introduced the term "molecular manufacturing," which he defined as the *"chemical synthesis of complex structures by mechanically positioning reactive molecules, not by manipulating individual atoms"*. Drexler described nanofactories in which nanomachines (resembling molecular assemblers, or industrial robot arms) combine molecules to build larger atomically precise parts. These parts, in turn, can be assembled by positioning mechanisms of assorted sizes to build macroscopic (visible) but still atomically-precise products. The concept is that a functioning nanofactory will create virtually any product at the cost of only the input raw material and energy.

"Nanomanufacturing" refers to the production of structures "bottom up" from nanoparticles (materials at the nanoscale of 10-9 meters) or "top down" in steps for high levels of precision. Unlike molecular manufacturing, it doesn't necessarily require chemical synthesis. Nanoparticles provide numerous possibilities for applications in nanotechnology due to their amazing properties. However, realizing their potential versatility requires assembly of nanoparticles in regular patterns on surfaces and at interfaces. Assembling nanoparticles generates new nanostructures, which in turn have unforeseen collective, intrinsic physical properties. These properties can be exploited for multipurpose applications in nanoelectronics, spintronics, sensors, and so forth.

A team led by Dr. Ting Xu at the U.S. Department of Energy's Lawrence Berkeley National Laboratory made an important advance towards this nanotechnology goal. They found a simple and yet powerful way to induce nanoparticles to assemble themselves into complex arrays. By adding specific types of small molecules to mixtures of nanoparticles and polymers, Dr. Xu's group directed the self-assembly of the nanoparticles into arrays of one, two and three dimensions with no chemical modification of either the nanoparticles or the block copolymers. In addition, they found that the application of external stimuli – light and/or heat – can be used to further direct the assemblies of nanoparticles for even finer and more complex structural details.

Small as they are, nanoparticles are essentially all surface. According to the BERKELEY LAB News Release on Dr. Xu's research, any process that modifies the surface of a nanoparticle can profoundly change the properties of that particle. Precisely arranging these nanoparticles is critical to tailoring the macroscopic properties during nanoparticle assembly. While chemical DNA can be used to induce self-assembly of nanoparticles with a high degree of precision, it only works well for organized arrays that are limited in size – it is impractical for large-scale fabrication. Dr. Xu's approach is to use block copolymers – long sequences or blocks of one type of monomer molecule bound to blocks of another type of monomer molecule. Like soldiers lining up in formation, the block copolymers assemble at densities of 10 trillion bits per square inch. Dr. Xu's technique promises to revolutionize the data storage industry, eventually leading to the contents of hundreds of DVDs — or its equivalent — fitting into a space the size of a thumbnail.

By adding specific types of small molecules to mixtures of nanoparticles and polymers, Dr. Xu's group directed the self-assembly of the nanoparticles. Dr. Xu, an assistant professor of materials science and engineering and of chemistry at UC Berkeley, is being honored as one of the "Brilliant 10" young researchers in the November 2009 issue of POPULAR SCIENCE. Her group is now working on applying the nanoparticle self-assembly technique to paper-thin, printable solar cells, and ultra-small electronic devices. "We've advanced the technique to make the nanocomposites responsive to light, which could enable the development of photovoltaic cells that are more energy efficient," says Xu. The nanofactories of tomorrow will likely require both molecular manufacturing as envisioned by Dr. Drexler using chemical synthesis and nanomanufacturing techniques like Dr. Xu's. Nanoparticles can now be induced to self-assemble non-chemically using block copolymers in regular patterns on surfaces and at interfaces to provide better data storage, solar cells, and tiny electronics.

Table 1. Various DNA structures (1D wires and ribbons, 1D tubes, 2D lattices, and 3D polygons).

1D Wire and Ribbon	Duplex DNA [24]	3-Helix Bundle [25] (1D)	TX Ribbon [26]	Rhombus-like Track [27]	4 × 4 Track [28]	Zig-Zag Ribbon [29]	Single Strand Tile Ribbon [30]
Schematic diagram							
Image							

1D Tube	TX Tube [31]	4 × 4 Tube [32]	DX Tube [33]	Chiral DX Tube [34]	6-Helix Bundle [35]	Single Strand Tile Tube [30]
Schematic diagram						
Image						

2D Lattice	DX Lattice [36]	4 × 4 Grid [32]	Triangle Lattice [38]	Symmetric 4 × 4 Grid [40]	3 Point-Star Lattice [41]	3-Helix Bundle [25] (2D)	Multiple-Helix Lattice [42]
Schematic diagram							
Image							

3D Polygon	Octahedron [43]	Tetrahedron [44]	Tetrahedron [45]	Dodecahedron [45]	Bucky Ball [45]
Schematic diagram					
Image					

DNA Tiles

A family of DNA tiles was developed and identified as DX tiles with a vast number of applications in DNA nanostructure fabrication. These tiles can form large 2D lattices and can be viewed by AFM (Atomic Force Microscopy). Later, other DNA tiles were also developed to provide for more complex strand topology and interconnections, like a family of DNA tiles known as TX tiles made by three DNA helices linked by four crossover junctions. These DNA tiles are the fundamental unit for the fabrication of a great number of DNA nanostructures and other applications. Standard ladder systems of DNA display a banding structure. These structures are now of great attention for semiconductor chip manufacturers. They are interested to use them for controlling tiny features on their chips. The DNA bands are unique from the banding found in other molecules, like collagen; here, a single piece of DNA can form ridges along the length of the crystal.

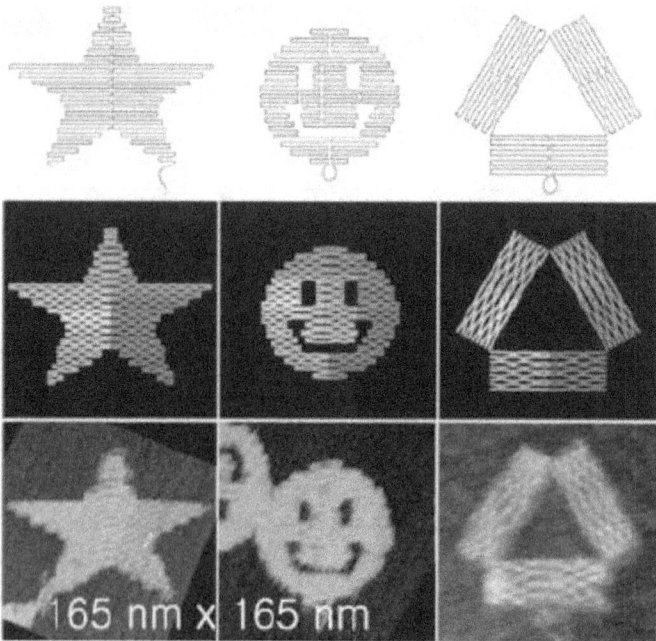

Fig. 21. Examples of some complex DNA structures by DNA origami.[57]

Computing with DNA

DNA sequence has an informational role in all different applications. Computation is possible with the information stored in DNA, was a big question. Computation may be defined as, "the process of getting an output from a number of inputs, using elementary instructions". A computer determines the sum of numbers in binary format, using the principles of Boolean logic. DNA is a striking medium because of its high information concentration where the accessible instructions rely on the context of the computation. Research in DNA computation, launched by Adleman, has opened the door for experimental study of the programmable biochemical reactions. He recognized that the density of information storage in DNA, in connection with the parallelism of chemical reactions, may be used to perform huge mathematical computations and so he created the field of DNA-based computers. Statistics shows that 1 mL of solution may have 10 bits of information encoded in DNA molecules and that chemical reactions (such as the activity of restriction enzymes, ligases, polymerases, or simple hybridization itself) can function on this information in parallel. The questions of what biochemical operations are best for molecular computation, how best to use them, whether they can be performed with sufficiently low errors and what specific applications are likely to be of practical interest, are all still open.

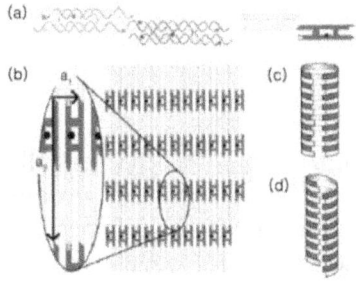

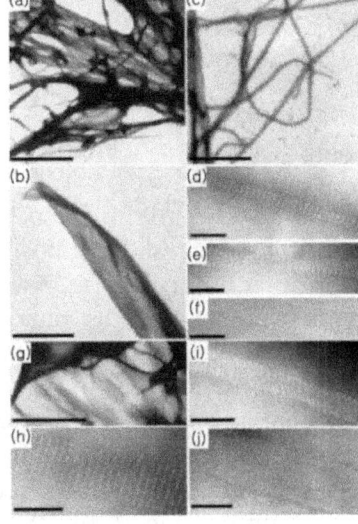

Fig. 10. Self-assembly of DNA tiles into sheets and tubes. (a) Structure of DX tiles. The 6 nt single-stranded sticky-ends on the α tile are complementary to those on the β tile. (b) α and β tiles tessellate to form extended two-dimensional arrays. 2D sheets can fold and close upon themselves to form tubes, producing either (c) alternating rings or (d) nested helices of α and β tiles.[34]

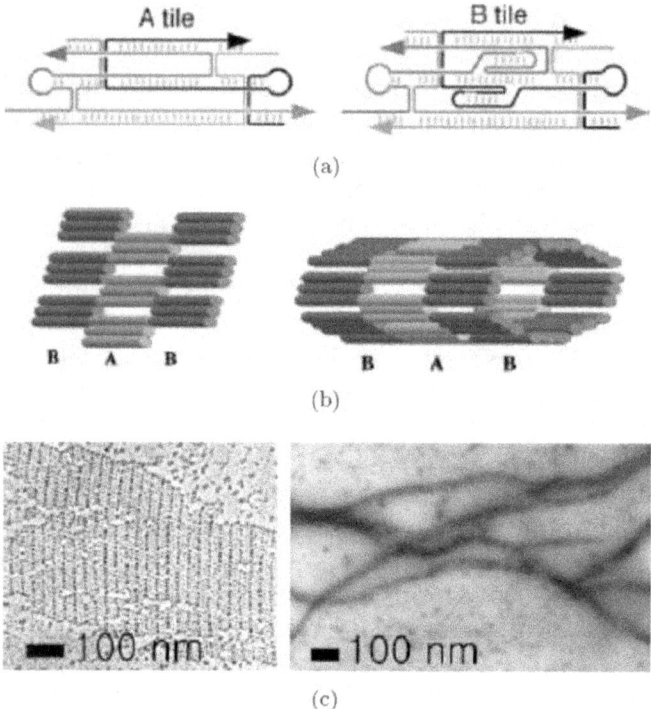

Fig. 7. (a) Schematics showing the strand traces through the two tile types, A and B, used in the constructions. (b) A TX flat lattice and a TX tube built from A and B tiles. (c) TEM images of platinum rotary-shadowed TX lattices and negative-stained TX nanotubes.[31]

Molecular diagnostics

DNA nanotechnology enables diagnosis at the single cell and molecule levels, and some may also be integrated in current molecular diagnostic methods, such as biochips. Nanoparticles, such as gold nanoparticles and quantum dots, are the most extensively used, but various other nanotechnological devices for manipulation at the nanoscale as well as nanobiosensors are also compete for potential clinical applications. Molecular diagnostics may assist to identify different disease markers, help the diagnosis of some cancers and immunodeficiency related diseases. Diagnostic methods may also identify allergens and pollutants. Highly sensitive detection permits more efficient screening at blood banks through the pooling of samples, and it facilitates the early disease diagnosis when treatment might be more promising and also in pediatrics, when sample size is inevitably very small.

Current DNA-based molecular diagnostics methods such as PCR (polymerase chain reaction) involved many cumbersome steps. A new streamlined method was developed that accomplishes highly sensitive detection at attomolar concentrations without the need for PCR amplification and is called bio-barcode amplification or BCA.

The target protein to be detected is "sandwiched" between two probes: a tagging probe and a capture probe. Both probes have antibodies that bind to the target protein such that the protein is sandwiched between the probes. BCA has been successfully applied for the diagnosis of prostate-specific antigen (an indicator of prostate cancer and breast cancer), amyloid-β-derived diffusable ligands (ADDLs; indicators of Alzheimer disease), and interleukin (a cytokine protein that is involved in inflammation and immune responses in humans).

Drug delivery
Among the suggested applications for structural DNA nanotechnology in the future is the packaging and delivery of nucleic acid, protein or small molecular drugs to particular locations within organisms, tissues or cells. DNA nanostructures would be logical vehicles for nucleic acid drugs (such as mRNA, miRNA, siRNA, or DNA), because they could be directly folded into the nanostructure using well-understood base-pairing associations. It also appears possible to organize and deliver bioactive protein enzymes to desired locations using DNA nanostructures as "rafts" upon which to organize targeting and therapeutic proteins. Some progress toward this goal has been reported.

Molecular Robotics
A diverse set of molecular machines, motors, walkers, and crawlers based on dynamic structural rearrangements in DNA nanostructures have been constructed, described, and reviewed elsewhere. This topic offers a wide range of potential applications for DNA-based nanotechnology constructions.

Nanoelectronics Components Organization

DNA may be used for the organization of nanoparticles and single-walled carbon nanotubes (CNTs). CNTs are intended to be use in tiny electronic transistors that may be formed into nanocircuits. There are a lot of challenges to build a circuit from CNTs, including controlling the layout of the component CNTs and placing the wires to connect the components according to a specific pattern. DNA tags offer great promise to meet these challenges. DNA tags that may be attached to CNTs can direct their assembly by binding a tag with its complement.

DNA strands have been used to spatially place CNTs, making it potentially able to assemble into circuits. Also ssDNA may be extended across a surface to attach two electrodes by binding the ends of the ssDNA with complementary strands that are attached to the electrodes. Metal (like silver, for example) can then be deposited along the extended DNA strand, to form a nanowire that connects the electrodes.

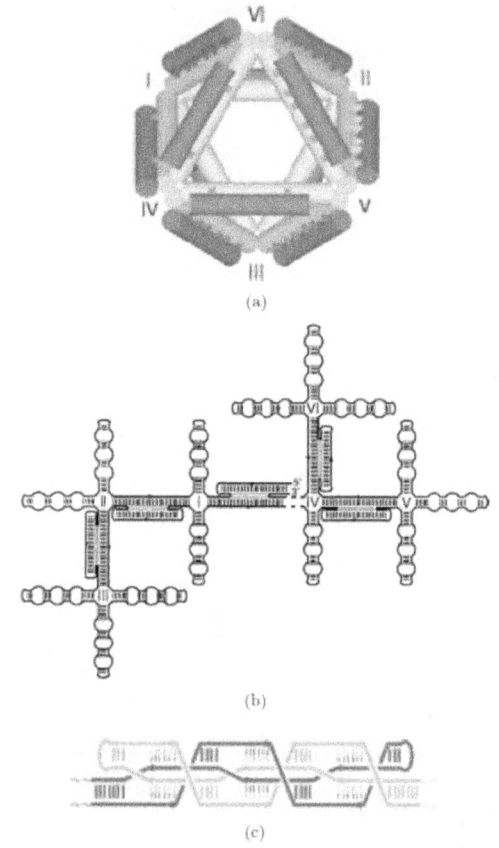

Fig. 22. Design of the DNA octahedron. (a) 3D structure involving twelve struts (octahedron edges) connected by six flexible joints (octahedron vertices). (b) Secondary structure of the branched-tree folding. (c) A schematic of a PX strut.[43]

Three-dimensional (3D) structure by self-assembly requires stiff building blocks. DNA is an ideal material for nanofabrication of rigid compositions because the assembly can be controlled by base-pairing and is comparatively cheaper and its implementation is simple. An early 3D DNA nanostructure was described, design and synthesis of a 1,669-nucleotide, single-stranded DNA molecule was reported; amplified by polymerases and that folds into an octahedron structure by a simple denaturation–renaturation procedure in the presence of five 40-mer synthetic oligodeoxynucleotides.

Using cryo-electron microscopy, the authors confirmed that the DNA strands fold successfully with 12 struts or edges joined at six four-way junctions to form hollow octahedra approximately 22 nm in diameter. The base-pair sequence of individual struts is not repetitive in a given octahedron, thus each strut is entirely addressable by the appropriate sequence-specific DNA binder. The DNA octahedron consists of five double crossover (DX) struts and seven paranemiccrossover (PX) struts, connected at six four-way junctions (Fig. 22(a)).

DX and PX motifs have both been designed as pairs of strut of the octahedron contributes to one double helix to a "core" layer and the other to a concentric "peripheral" layer.

The four-way junctions here tie only the core-layer double helices. Each four-way junction displays on its concave face the minor grooves of its four proximally surrounding base pairs. All four strands at all six junctions have two unpaired thymidine remainder at the crossover point to give some flexibility for the assembly.

The core-layer double helix of each of the twelve struts includes 40 base pairs, consequent to almost four turns of DNA and a length of 14 nm. For the eleventh struts, the peripheral-layer double helix have 30 or 32 base pairs and is restricted at both helical ends by a hairpin loop of four thymidine residues. The twelfth strut is a little longer including 35 base pairs and is checked at only one end.

DX motifs are twice as stiff as standard duplex DNA. The folded octahedron with all twelve struts together is expected to be a highly stiff object. The DNA octahedron structure was exposed by using cryo-electron microscopy and single particle reconstruction techniques.63,64 Figure 23(a) shows the cryo-electron micrograph, presenting numerous octahedral-shaped objects of the expected size. Figure 23(b) demonstrates a three-dimensional drawing of the structure of the DNA octahedron, reconstructed from 961 particles. Figure 23(c) explains the images of the representative particles, each comes with a corresponding projection of the computed map or drawing. The chiral assembly of DNA rigid structure for 3D nanofabrication44 was done in 2005. They produced a class of DNA tetrahedra that self-assemble in a single step and in only a few seconds.

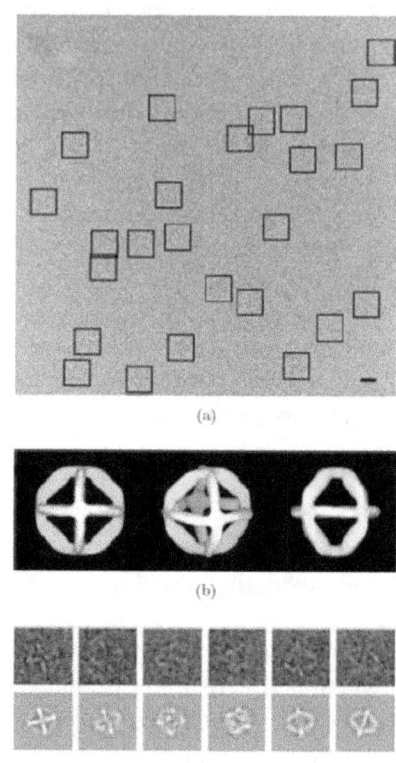

Fig. 23. DNA octahedron structure cryo-electron micros-copy. (a) Cryo-electron micrograph, with 25 individual DNA particles boxed. Scale bar, 20 nm. (b) 3D map generated from multiple particle reconstruction of the DNA octahedron. (c) Raw images of individual particles and the corresponding projections of the 3D map.43

Their flexibility proved them as building blocks for 3D nanofabrication by assembling one regular and nine different asymmetrical tetrahedral and by linking them with programmable DNA linkers. They used AFM to image the tertiary structure of individual tetrahedra and to reveal their rigidity. The DNA tetrahedron was designed to be mechanically robust. It was made by rigid triangles of DNA helices covalently joined at the vertices. The four-component oligonucleotides each run about one face and hybridize to form the double helical edges. Four edges contain nicks (i.e., breaks in the DNA backbone) where the ends of an oligonucleotide get assembled. In contrast to the challenging fabrication of DNA cubes and octahera, the synthesis of tetrahedra is really simple. All four oligonucleotides combined in equimolar concentration in hybridization buffer at 95°C and then cooled to 4°C in 30 seconds.

In conclusion, structural DNA nanotechnology has already been used to demonstrate the design and construction of a fascinating array of nanoscale objects and structures. DNA has been used to create the largest, artificial, fully-addressable, self-assembled, nanoscale objects known to science.

The prospects for commercial, medical, governmental, and other applications of DNA-based nanotechnology in the very near future are essentially inevitable.

6 ARTIFICIAL INTELLIGENCE AND INFORMATICS

*"**Artificial Intelligence** is a field of study which examines systems related to life's processes, evolution, biology, language and cognition (to name a few) using complex computational models with the goal of understanding information processing using the ever expanding digital environment."* - Indiana University Bloomington

In other words artificial intelligence is the science and engineering of making intelligent machines, especially intelligent computer programs. It is related to the similar task of using computers to understand human intelligence, but AI does not have to confine itself to methods that are biologically observable. At this point, it would be appropriate to understand what "intelligence" is from a scientific point of view. Intelligence is simply the computational part of the ability to achieve goals in the world. Varying kinds and degrees of intelligence occur in people, many animals and some machines.

There isn't really a solid definition of intelligence that doesn't depend on relating it to human intelligence. The problem is that we cannot yet characterize in general what kinds of computational procedures we want to call intelligent. We understand some of the mechanisms of intelligence and not others.

Intelligence is not a single thing that one can ask a yes or no question of, for example; "*Is this machine intelligent or not?*" Intelligence involves mechanisms, and AI research has discovered how to make computers carry out some of them and not others. If doing a task requires only mechanisms that are well understood today, computer programs can give very impressive performances on these tasks. Such programs should be considered "somewhat intelligent".

AI isn't always about simulating human intelligence. Sometimes... but not always or even usually. On the one hand, we can learn something about how to make machines solve problems by observing other people or just by observing our own methods. On the other hand, most work in AI involves studying the problems the world presents to intelligence rather than studying people or animals. AI researchers are free to use methods that are not observed in people or that involve much more computing than people can do.

Arthur R. Jensen a leading researcher in human intelligence, suggests "*as a heuristic hypothesis*" that all normal humans have the same intellectual mechanisms and that differences in intelligence are related to "*quantitative biochemical and physiological conditions*". Things such as; speed, short term memory, and the ability to form accurate and retrievable long term memories. Whether or not Jensen is right about human intelligence, the situation in AI today is the reverse.

Computer programs have plenty of speed and memory but their abilities correspond to the intellectual mechanisms that program designers understand well enough to put in programs. Some abilities that children normally don't develop till they are teenagers may be in, and some abilities possessed by two year olds are still out. The matter is further complicated by the fact that the cognitive sciences still have not succeeded in determining exactly what the human abilities are. Very likely the organization of the intellectual mechanisms for AI can usefully be different from that in people.

Whenever people do better than computers on some task or computers use a lot of computation to do as well as people, this demonstrates that the program designers lack understanding of the intellectual mechanisms required to do the task efficiently.

Initially, AI became popular after WWII, when a number of people independently started to work on intelligent machines. The English mathematician Alan Turing may have been the first. He gave a lecture on it in 1947. He also may have been the first to decide that AI was best researched by programming computers rather than by building machines. By the late 1950s, there were many researchers on AI, and most of them were basing their work on programming computers.

Some researchers currently have the objective of merging the human mind with computers, but the human mind has a lot of peculiarities, and imitating it would not be as beneficial as integration into it.

Alan Turing's 1950 article *Computing Machinery and Intelligence* discussed conditions for considering a machine to be intelligent. He argued that if the machine could successfully pretend to be human to a knowledgeable observer then you certainly should consider it intelligent. This test would satisfy most people but not all philosophers. The observer could interact with the machine and a human by teletype (to avoid requiring that the machine imitate the appearance or voice of the person), and the human would try to persuade the observer that it was human and the machine would try to fool the observer.

The Turing test is a one-sided test. A machine that passes the test should certainly be considered intelligent, but a machine could still be considered intelligent without knowing enough about humans to imitate a human.

Daniel Dennett's book *Brainchildren* has an excellent discussion of the Turing test and the various partial Turing tests that have been implemented, i.e. with restrictions on the observer's knowledge of AI and the subject matter of questioning. It turns out that some people are easily led into believing that a rather dumb program is intelligent.

AI does however, aim at human-level intelligence. The ultimate effort is to make computer programs that can solve problems and achieve goals in the world as well as humans. However, many people involved in particular research areas are much less ambitious.

A few people think that human-level intelligence can be achieved by writing large numbers of programs of the kind people are now writing and assembling vast knowledge bases of facts in the languages now used for expressing knowledge.

However, most AI researchers believe that new fundamental ideas are required, and therefore it cannot be predicted when human-level intelligence will be achieved.

Computers can be programmed to simulate any kind of machine. Many researchers invented non-computer machines, hoping that they would be intelligent in different ways than the computer programs could be. However, they usually simulate their invented machines on a computer and come to doubt that the new machine is worth building. Because many billions of dollars that have been spent in making computers faster and faster, another kind of machine would have to be very fast to perform better than a program on a computer simulating the machine.

Some people think much faster computers are required as well as new ideas. It is interesting when we contemplate the possibility that the computers of 30 years ago were fast enough if only we knew how to program them. Of course, quite apart from the ambitions of AI researchers, computers will keep getting faster.

Machines with many processors are much faster than single processors can be. Parallelism itself presents no advantages, and parallel machines are somewhat awkward to program. When extreme speed is required, it is necessary to face this awkwardness.

Perhaps we could just make an AI machine that simply learns on its own. Actually the idea of making a "child machine" that could improve by reading and learning from experience has been proposed many times, starting in the 1940s. Eventually, it will be made to work. However, AI programs haven't yet reached the level of being able to learn much of what a child learns from physical experience. Nor do present programs understand language well enough to learn much by reading.

Another popular benchmark in AI, is the computer that can play chess against a human opponent. Alexander Kronrod, a Russian AI researcher, said "Chess is the *Drosophila* of AI." He was making an analogy with geneticists' use of that fruit fly to study inheritance. Playing chess requires certain intellectual mechanisms and not others. Chess programs now play at grandmaster level, but they do it with limited intellectual mechanisms compared to those used by a human chess player, substituting large amounts of computation for understanding. Once we understand these mechanisms better, we can build human-level chess programs that do far less computation than do present programs.

Unfortunately, the competitive and commercial aspects of making computers play chess have taken precedence over using chess as a scientific domain. It is as if the geneticists after 1910 had organized fruit fly races and concentrated their efforts on breeding fruit flies that could win these races.

The Chinese and Japanese game of *Go* is also a board game in which the players take turns moving. *Go* exposes the weakness of our present understanding of the intellectual mechanisms involved in human game playing. *Go* programs are very bad players, in spite of considerable effort (not as much as for chess). The problem seems to be that a position in *Go* has to be divided mentally into a collection of sub-positions which are first analyzed separately followed by an analysis of their interaction. Humans use this in chess also, but chess programs consider the position as a whole. Chess programs compensate for the lack of this intellectual mechanism by doing thousands or, in the case of Deep Blue, many millions of times as much computation.

Sooner or later, AI research will overcome this disreputable weakness.

Some people say that AI is a bad idea, just think of popular films such as Terminator and The Matrix where forms of AI have essentially caused the literal annihilation and/or enslavement of the human race. The philosopher John Searle says that the idea of a non-biological machine being intelligent is incoherent.

The philosopher Hubert Dreyfus says that AI is impossible. The computer scientist Joseph Weizenbaum says the idea is obscene, anti-human and immoral. Various people have said that since artificial intelligence hasn't reached human level by now, it must be impossible. Still other people are disappointed that companies they invested in went bankrupt.

In the 1930s mathematical logicians, especially Kurt Gödel and Alan Turing, established that there did not exist algorithms that were guaranteed to solve all problems in certain important mathematical domains. Whether a sentence of first order logic is a theorem is one example, and whether a polynomial equations in several variables has integer solutions is another.

Humans solve problems in these domains all the time, and this has been offered as an argument (usually with some decorations) that computers are intrinsically incapable of doing what people do. Roger Penrose claims this.

However, people can't guarantee to solve *arbitrary* problems in these domains either.

In the 1960s computer scientists, Steve Cook and Richard Karp developed the theory of NP-complete problem domains. Problems in these domains are solvable, but seem to take exponential amounts of time in comparison to the size of the problem. Which sentences of propositional calculus are satisfiable is a basic example of an NP-complete problem domain. Humans often solve problems in NP-complete domains in times much shorter than is guaranteed by the general algorithms, but can't solve them quickly in general.

What is important for AI is to have algorithms as capable as people at solving problems. The identification of subdomains for which good algorithms exist is important, but a lot of AI problem solvers are not associated with readily identified subdomains.

The theory of the difficulty of general classes of problems is called *computational complexity.* So far this theory hasn't interacted with AI as much as might have been hoped. Success in problem solving by humans and by AI programs seems to rely on properties of problems and problem solving methods that the neither the complexity researchers nor the AI community have been able to identify precisely.

Algorithmic complexity theory as developed by Solomonoff, Kolmogorov and Chaitin (independently of one another) is also relevant. It defines the complexity of a symbolic object as the length of the shortest program that will generate it. Proving that a candidate program is the shortest or close to the shortest is an unsolvable problem, but representing objects by short programs that generate them should sometimes be illuminating even when you can't prove that the program is the shortest.

Branches of AI

What kinds of things can today's AI accomplish? What are researchers currently working on as far as AI is concerned?

I have attempted to compile a complete list but some branches are surely missing, because no-one has identified them yet. Some of these may be regarded as concepts or topics rather than full branches.

Logical AI

What a program knows about the world in general the facts of the specific situation in which it must act, and its goals are all represented by sentences of some mathematical logical language. The program decides what to do by inferring that certain actions are appropriate for achieving its goals.

S*earch*

AI programs often examine large numbers of possibilities, e.g. moves in a chess game or inferences by a theorem proving program. Discoveries are continually made about how to do this more efficiently in various domains.

Pattern recognition

When a program makes observations of some kind, it is often programmed to compare what it sees with a pattern. For example, a vision program may try to match a pattern of eyes and a nose in a scene in order to find a face. More complex patterns, e.g. in a natural language text, in a chess position, or in the history of some event are also studied. These more complex patterns require quite different methods than do the simple patterns that have been studied the most.

Representation

Facts about the world have to be represented in some way. Usually languages of mathematical logic are used.

Inference

From some facts, others can be inferred. Mathematical logical deduction is adequate for some purposes, but new methods of *non-monotonic* inference have been added to logic since the 1970s. The simplest kind of non-monotonic reasoning is default reasoning in which a conclusion is to be inferred by default, but the conclusion can be withdrawn if there is evidence to the contrary. For example, when we hear of a bird, we man infer that it can fly, but this conclusion can be reversed when we hear that it is a penguin. It is the possibility that a conclusion may have to be withdrawn that constitutes the non-monotonic character of the reasoning. Ordinary logical reasoning is monotonic in that the set of conclusions that can the drawn from a set of premises is a monotonic increasing function of the premises. Circumscription is another form of non-monotonic reasoning.

Common sense knowledge and reasoning

This is the area in which AI is farthest from human-level, in spite of the fact that it has been an active research area since the 1950s. While there has been considerable progress, e.g. in developing systems of *non-monotonic reasoning* and theories of action, yet more new ideas are needed. The Cyc system contains a large but spotty collection of common sense facts.

Learning from experience

Programs do that. The approaches to AI based on *connectionism* and *neural nets* specialize in that. There is also learning of laws expressed in logic. Programs can only learn what facts or behaviors their formalisms can represent, and unfortunately learning systems are almost all based on very limited abilities to represent information.

Planning

Planning programs start with general facts about the world (especially facts about the effects of actions), facts about the particular situation and a statement of a goal. From these, they generate a strategy for achieving the goal. In the most common cases, the strategy is just a sequence of actions.

Epistemology

This is a study of the kinds of knowledge that are required for solving problems in the world.

Ontology

Ontology is the study of the kinds of things that exist. In AI, the programs and sentences deal with various kinds of objects, and we study what these kinds are and what their basic properties are. Emphasis on ontology begins in the 1990s.

Heuristics

A heuristic is a way of trying to discover something or an idea imbedded in a program. The term is used variously in AI. *Heuristic functions* are used in some approaches to search to measure how far a node in a search tree seems to be from a goal. *Heuristic predicates* that compare two nodes in a search tree to see if one is better than the other, i.e. constitutes an advance toward the goal, may be more useful.

Genetic programming

Genetic programming is a technique for getting programs to solve a task by mating random Lisp programs and selecting fittest in millions of generations.

Applications of AI

There are many things that AI can be used for, here are a few examples that have already been developed, and some that we are sure to see implemented soon.

Game playing

You can buy machines that can play master level chess for a few hundred dollars. There is some AI in them, but they play well against people mainly through brute force computation--looking at hundreds of thousands of positions. To beat a world champion by brute force and known reliable heuristics requires being able to look at 200 million positions per second.

Speech recognition

In the 1990s, computer speech recognition reached a practical level for limited purposes. Thus United Airlines has replaced its keyboard tree for flight information by a system using speech recognition of flight numbers and city names. It is quite convenient. On the the other hand, while it is possible to instruct some computers using speech, most users have gone back to the keyboard and the mouse as still more convenient.

Understanding natural language

Just getting a sequence of words into a computer is not enough. Parsing sentences is not enough either. The computer has to be provided with an understanding of the domain the text is about, and this is presently possible only for very limited domains.

Computer vision

The world is composed of three-dimensional objects, but the inputs to the human eye and computers' TV cameras are two dimensional. Some useful programs can work solely in two dimensions, but full computer vision requires partial three-dimensional information that is not just a set of two-dimensional views. At present there are only limited ways of representing three-dimensional information directly, and they are not as good as what humans evidently use.

Expert systems

A ``knowledge engineer" interviews experts in a certain domain and tries to embody their knowledge in a computer program for carrying out some task. How well this works depends on whether the intellectual mechanisms required for the task are within the present state of AI. When this turned out not to be so, there were many disappointing results. One of the first expert systems was MYCIN in 1974, which diagnosed bacterial infections of the blood and suggested treatments. It did better than medical students or practicing doctors, provided its limitations were observed. Namely, its ontology included bacteria, symptoms, and treatments and did not include patients, doctors, hospitals, death, recovery, and events occurring in time. Its interactions depended on a single patient being considered. Since the experts consulted by the knowledge engineers knew about patients, doctors, death, recovery, etc., it is clear that the knowledge engineers forced what the experts told them into a predetermined framework. In the present state of AI, this has to be true. The usefulness of current expert systems depends on their users having common sense.

Heuristic classification

One of the most feasible kinds of expert system given the present knowledge of AI is to put some information in one of a fixed set of categories using several sources of information. An example is advising whether to accept a proposed credit card purchase. Information is available about the owner of the credit card, his record of payment and also about the item he is buying and about the establishment from which he is buying it (e.g., about whether there have been previous credit card frauds at this establishment).

~

AI research has both theoretical and experimental sides. The experimental side has both basic and applied aspects.

There are two main lines of research. One is biological, based on the idea that since humans are intelligent, AI should study humans and imitate their psychology or physiology. The other is phenomenal, based on studying and formalizing common sense facts about the world and the problems that the world presents to the achievement of goals. The two approaches interact to some extent, and both should eventually succeed. It is a race, but both racers seem to be walking.

AI has many relations with philosophy, especially modern analytic philosophy. Both study mind, and both study common sense.

Informatics

Informatics is the study of the structure, behavior, and interactions of natural and engineered computational systems; and has also recently become the replacement word of choice for "AI" or "Artificial Intelligence".

The central focus of Informatics is the transformation of information - whether by computation or communication, whether by organisms or artifacts. Understanding informational phenomena - such as computation, cognition, and communication - enables technological advances.

Informatics also informs and is informed by other disciplines, such as Mathematics, Electronics, Biology, Linguistics and Psychology. Consequently, Informatics provides a link between disciplines with their own methodologies and perspectives and brings together a common scientific paradigm.

Informatics studies the representation, processing, and communication of information in natural and engineered systems.

It has computational, cognitive and social aspects. The central notion is the transformation of information - whether by computation or communication, whether by organisms or artifacts. Understanding informational phenomena - such as computation, cognition, and communication - enables technological advances.

In turn, technological progress prompts scientific enquiry. The science of information and the engineering of information systems develop hand-in-hand. Informatics is the emerging discipline that combines the two.

In natural and artificial systems, information is carried at many levels, ranging, for example, from biological molecules and electronic devices through nervous systems and computers and on to societies and large-scale distributed systems. It is characteristic that information carried at higher levels is represented by informational processes at lower levels. Each of these levels is the proper object of study for some discipline of science or engineering. Informatics aims to develop and apply firm theoretical and mathematical foundations for the features that are common to all computational systems.

The Scope Of Informatics

In its attempts to account for phenomena, science progresses by defining, developing, criticizing and refining new concepts. Informatics is developing its own fundamental concepts of communication, knowledge, data, interaction and information, and relating them to such phenomena as computation, thought, and language.

Informatics has many aspects, and encompasses a number of existing academic disciplines - Artificial Intelligence, Cognitive Science and Computer Science. Each takes part of Informatics as its natural domain: in broad terms, Cognitive Science concerns the study of natural systems; Computer Science concerns the analysis of computation, and design of computing systems; Artificial Intelligence plays a connecting role, designing systems which emulate those found in nature. Informatics also informs and is informed by other disciplines, such as Mathematics, Electronics, Biology, Linguistics and Psychology. Thus Informatics provides a link between disciplines with their own methodologies and perspectives, bringing together a common scientific paradigm, common engineering methods and a pervasive stimulus from technological development and practical application.

Three of the truly fundamental questions of Science are: "What is matter?", "What is life?" and "What is mind?". The physical and biological sciences concern the first two. The emerging science of Informatics contributes to our understanding of the latter two by providing a basis for the study of organization and process in biological and cognitive systems.

Progress can best be made by means of strong links with the existing disciplines devoted to particular aspects of these questions.

Computational Systems

Computational systems, whether natural or engineered, are distinguished by their great complexity, as regards both their internal structure and behavior, and their rich interaction with the environment. Informatics seeks to understand and to construct (or reconstruct) such systems, using analytic, experimental and engineering methodologies. The mixture of observation, theory and practice will vary between natural and artificial systems.

In natural systems, the object is to understand the structure and behavior of a given computational system. The theoretical concepts underlying natural systems ultimately are built on observation and are themselves used to predict new observations. For engineered systems, the object is to build a system that performs a given informational function. The theoretical concepts underlying engineered systems are intended to secure their correct and efficient design and operation.

Informatics provides an enormous range of problems and opportunities. One challenge is to determine how far, and in what circumstances, theories of information processing in artificial devices can be applied to natural systems. A second challenge is to determine how far principles derived from natural systems are applicable to the development of new kinds of engineered systems. A third challenge is to explore the many ways in which artificial information systems can help to solve problems facing mankind and help to improve the quality of life for all living things. One can also consider systems of mixed character; a question of longer term interest may be to what extent it is helpful to maintain the distinction between natural and engineered systems.

Computational systems, whether natural or engineered, are distinguished by their great complexity. Informatics seeks to understand and to construct (or reconstruct) such systems, using analytic, experimental and engineering methodologies.

By studying computational systems, Informatics seeks to address the following challenges: Determining how far, and in what circumstances, theories of information processing in artificial devices can be applied to natural systems. Determining how far principles derived from natural systems are applicable to the development of new kinds of engineered systems. Exploring the many ways in which artificial information systems can help to solve problems facing mankind and help to improve the quality of life for all living things.

Environmental Informatics may be considered as the combination of software and environmental engineering methods and tools for the creation of a new "knowledge-paradigm" towards serving environmental engineering needs. EI is therefore an integrator of science, methods, and techniques and not just the result of using information and software technology methods and tools for solving environmental problems. To this end, EI is a "bridging discipline" dealing with the development and application of Information Society Technologies (IST) systems in the environmental domain. Being a relatively new discipline, EI aims to furnish systems engineers attempting to manage and direct information flow regarding the environmental domain to decision makers and the public with time-proven methodologies that have evolved from the application of the Software Engineering discipline. To this end, Artificial Intelligence may very well serve the needs of environmental engineering and management problems, by providing with methods for knowledge extraction and key parameter identification for better understanding and analysis of environmental problems. Moreover, AI may provide with various methods and algorithms for simulating, modeling, and forecasting the behavior of environmental systems. This is of particular importance in the case of quality of life oriented problems, like in the case of air pollutant concentrations in the atmospheric environment. Environmental knowledge engineering and extraction System analysis and modeling, Environmental quality monitoring and forecasting, Environmental information management, Web technologies and AI methods for the environmental sector, Quality of life services, Urban pollution problems, Combined EI and AI.

Researchers are focusing on areas that include: The evolution and analysis of dynamical nervous systems for model agents. Computational and theoretical biology, including models of metabolism, gene regulation and development, computer vision, inferring semantic meaning from images, applications of computational linguistics to the democratization of knowledge, computational models of our human concepts and categories, robot motion planning and control, semiautonomous robots, and integrating perception and planning, case-based reasoning, intelligent information systems, intelligent user interfaces, knowledge management, knowledge modeling, multimodal reasoning, multi-strategy learning, and introspective reasoning, complex systems, adaptive agents, modeling, simulation, artificial life, and complex (information, biological, and social) networks - the Web as a complex information network in which we leave abundant traces of our social and semantic activities, musical research includes accompaniment systems, computer generated musical analysis, musical signal processing, and modeling of musical interpretation, informational properties of natural and artificial systems which enable them to adapt and evolve, computational ecology and evolutionary trends in an information-theoretic measure of the complexity of neural structure and function.

Bioinformatics

Bioinformatics is the field of science in which biology, computer science, and information technology merge to form a single discipline. The ultimate goal of the field is to enable the discovery of new biological insights as well as to create a global perspective from which unifying principles in biology can be discerned. At the beginning of the "genomic revolution", a bioinformatics concern was the creation and maintenance of a database to store biological information, such as nucleotide and amino acid sequences. Development of this type of database involved not only design issues but the development of complex interfaces whereby researchers could both access existing data as well as submit new or revised data.

Ultimately, however, all of this information must be combined to form a comprehensive picture of normal cellular activities so that researchers may study how these activities are altered in different disease states. Therefore, the field of bioinformatics has evolved such that the most pressing task now involves the analysis and interpretation of various types of data, including nucleotide and amino acid sequences, protein domains, and protein structures. The actual process of analyzing and interpreting data is referred to as **computational biology**. Important sub-disciplines within bioinformatics and computational biology include: the development and implementation of tools that enable efficient access to, and use and management of, various types of information. And also the development of new algorithms (mathematical formulas) and statistics with which to assess relationships among members of large data sets, such as methods to locate a gene within a sequence, predict protein structure and/or function, and cluster protein sequences into families of related sequences.

Over the past few decades, major advances in the field of molecular biology, coupled with advances in genomic technologies, have led to an explosive growth in the biological information generated by the scientific community. This deluge of genomic information has, in turn, led to an absolute requirement for computerized databases to store, organize, and index the data and for specialized tools to view and analyze the data.

A **biological database** is a large, organized body of persistent data, usually associated with computerized software designed to update, query, and retrieve components of the data stored within the system. A simple database might be a single file containing many records, each of which includes the same set of information. For example, a record associated with a nucleotide sequence database typically contains information such as contact name, the input sequence with a description of the type of molecule, the scientific name of the source organism from which it was isolated, and often, literature citations associated with the sequence.

For researchers to benefit from the data stored in a database, two additional requirements must be met; easy access to the information, and a method for extracting only that information needed to answer a specific biological question.

The rapidly emerging field of bioinformatics promises to lead to advances in understanding basic biological processes and, in turn, advances in the diagnosis, treatment, and prevention of many genetic diseases.

Bioinformatics has transformed the discipline of biology from a purely lab-based science to an information science as well. Increasingly, biological studies begin with a scientist conducting vast numbers of database and web site searches to formulate specific hypotheses or to design large-scale experiments. The implications behind this change, for both science and medicine, are staggering.

7 SYNTHETIC BIOLOGY

Synthetic Biology is a newer field of biology that aims at designing and building original biological systems. Whereas "genetic engineering" has been around for many decades already, synthetic biology extends its spirit to focus on *whole systems* of genes and gene products, rather than on *individual* genes. Thus, synthetic biology aims, for illustration, at adding or modifying biological functions to existing organisms or, in the future, creating new organisms with tailored properties. Synthetic biology will inevitably revolutionize how we conceptualize and approach the engineering of biological systems. The vision and applications of this emerging field will influence many other scientific and engineering disciplines, as well as affect various aspects of daily life and society.

The focus on systems as opposed to individual genes or pathways is shared by the contemporaneous discipline of systems biology, which analyzes biological organisms in their entirety. Synthetic biologists design and construct complex artificial biological systems using many insights discovered by systems biologists and share their holistic perspective.

Synthetic biology refers to both; The design and fabrication of biological components and systems that do not already exist in the natural world, and also the re-design and fabrication of existing biological systems.

"Synthetic biology," in the modern sense, means using engineering principles to create functional systems based on the molecular machines and regulatory circuits of living organisms found in nature. However, it also includes going beyond them to develop radically new systems. At the very least, synthetic biology represents a merger of molecular biology, genetic engineering and computer science – and given the scale of the components that it works with, it also qualifies as a form of nanotechnology.

What this means is that finally, after thousands of years of genetic manipulation by selective breeding; man has finally gained the ability to directly access the genetic code itself. The process for ordering synthetic DNA is as easy as placing an order via an online retailer. Yes, you can actually custom order any DNA sequence you like via the internet. The following is a testimonial from GenScript.com, a full-service online CRO:

"GenScript provides fast, professional protein synthesis services at very reasonable prices. By making it cost-effective to outsource protein production, GenScript has made it possible for my lab to focus on our own area of expertise and get more research done. The detailed planning, updates, and reports that GenScript provides all of the quality control that one could ask for. I strongly recommend GenScript its protein production service." — Scientist, Kansas State University, USA

Such contract research organizations, or CROs, charge large drug makers and other companies for conducting all stages of research -- from discovery of compounds that might be used as drugs, to animal studies and expensive clinical trials involving hundreds or thousands of patients with various ailments. Thanks to these type of private organizations, we not only need to now be concerned with an escaping new super-pathogen, but even the very ethics that these corporations follow. For example; what's wrong with the following quote from a 2009 Reuters article?

"With an abundant supply of primates, and less animal-rights advocacy, China has become a favorable destination for animal testing.", said Frank Zhang, chief executive of GenScript, a biology CRO with operations in the United States and China.

More and more synthetic biologists are engineering ever more complex artificial biological systems to investigate natural biological phenomena and for a variety of other (often not so benign) applications. The basis of synthetic biology are a relatively new engineering discipline, using research from the latest synthesis techniques and essentially replacing and even reinventing nature with man-made synthesized genetic materials. The fact that we are now essentially creating "life" makes it unique among all other existing engineering fields, and I believe that if we must continue with this technology, we should proceed with extreme caution.

The field of Synthetic Biology allows the design and construction of engineered cells with novel functions in a framework of an abstract hierarchy of biological devices, modules, cells, and multicellular systems.

The synthesized DNA sequence may be copied from nature, but the DNA itself is made by a machine. This makes it "synthetic". There currently are many sets that have been developed that can do different things; such as integrating the bioluminescent glow from a firefly, the smell of a banana, and any other result of DNA found in nature (and increasingly more commonly... not found in nature). As a matter of fact as this technology advances, the possibilities within it are simply astounding.

The raw material for synthesizing DNA is sugar. $25 of which will buy you enough to make a copy of every human genome on the planet. Your custom sequence information is fed to the DNA equivalent of an industrial laser printer. It receives your information, and out comes your custom DNA. It is then freeze-dried and shipped direct to your door. It currently only costs around $.40 per base pair, and it is getting cheaper all the time.

Engineers have already even assembled an open source catalog of over five thousand components called BioBricks. There is even an annual worldwide convention in which university students build new and more complex BioBricks, which they then string together and run inside a bacteria that we have all become more familiar with in recent years as there have been multiple "outbreaks" of; E.-coli, currently the most commonly (publicly) synthesized organism.

There are even new synthetic versions of the E.-coli bacteria that have been developed, such as; E. Chromi - Which includes a built-in sensitivity tuner and color generator which is programmed to turn one of five colors when it detects a certain concentration of an environmental toxin. There is also E. Coliroid - A bacterial system which switches on and off in response to red light, and acts like a bacterial Polaroid camera. These are just the results of groups that are "playing" with this technology like it's a toy... and because of this, we must ask; what are the groups with a lot more money and resources doing with this technology (that is not public)?

There are two types of synthetic biologists. The first group uses unnatural molecules to mimic natural molecules with the goal of creating artificial life. The second group uses natural molecules and assembles them into a system that acts unnaturally. In general, the stated goal is to solve problems that are not easily understood through analysis and observation alone and this is achieved by the manifestation of new models. So far, synthetic biology has produced diagnostic tools for diseases such as HIV and hepatitis viruses as well as devices from biomolecular parts with interesting functions.

The term "synthetic biology" was first used on genetically engineered bacteria that were created with recombinant DNA technology which was synonymous with bioengineering. Later the term "synthetic biology" was used as a means to redesign life which is an extension of biomimetic chemistry, where organic synthesis is used to generate artificial molecules that mimic natural molecules such as enzymes.

Synthetic biologists are trying to assemble unnatural components to support Darwinian evolution.

Recently, the engineering community has been seeking to extract components from the biological systems to test and confirm them as building units to be reassembled in a way that can mimic nature. In the engineering aspect of synthetic biology, the suitable parts are the ones that can contribute independently to the whole system so that the behavior of an assembly can be predicted. DNA consists of double-stranded anti-parallel strands each having four various nucleotides assembled from bases, sugars and phosphates which are made of carbon, nitrogen, oxygen, hydrogen and phosphorus atoms. In the Watson-Crick model, A pairs with T and G pairs with C although occasionally some diversity exists. This simplification doesn't exist in proteins. With analysis and observation alone, scientists convince themselves that the paradigms are the truth and if the data contradicts the theory, the data normally is discarded as an error, where synthesis encourages scientists to cross into the new land and define new theories. The same synthesis has long been used in chemistry such as chromatography. The combination of chemistry, biology and engineering can therefore create artificial Darwinian systems.

Synthetic biology studies how to build artificial biological systems for engineering applications, using many of the same tools and experimental techniques as systems biology. But the work is fundamentally an engineering application of biological science, rather than an attempt to do more science. The focus is often on ways of taking parts of natural biological systems, characterizing and simplifying them, and using them as a component of a highly unnatural, engineered, biological system.

Biologists are interested in synthetic biology because it provides a complementary perspective from which to consider, analyze, and ultimately understand the living world. Being able to design and build a system is also one very practical measure of understanding. Physicists, chemists and others are interested in synthetic biology as an approach with which to probe the behavior of molecules and their activity inside living cells. For example, differences between how a synthetic system is designed to behave and how it actually behaves can serve to highlight relevant intracellular physics. Engineers are interested in synthetic biology because the living world provides a seemingly rich yet largely unexplored medium for controlling and processing information, materials, and energy. Learning how to effectively harness the power of the living world will be a major engineering undertaking.

Scientists are working to create a general scientific and technical infrastructure that supports the design and synthesis of biological systems. Specifically they are working to specify and populate a set of standard parts that have well-defined performance characteristics and can be used (and re-used) to build biological systems (think of Lego building bricks), develop and incorporate design methods and tools into a integrated engineering environment, reverse engineer and re-design pre-existing biological parts and devices in order to expand the set of functions that they can access and program, and reverse engineer and re-design a 'simple' natural bacterium.

One might ask; why are they working to redesign bacterium? Bacteria are the simplest known objects from the natural world that are capable of replicating when provided with only simpler components (e.g., broth). Still, bacteria are far from simple. Bacteria also provide the basic environment in which synthetic biological systems exist and act (i.e., they are like the power supply and chassis of a computer). By re-designing/refactoring a simple living system, synthetic biologists hope to learn how to better couple (and decouple) our designed systems from their host environment.

Many technologies have the potential to be dangerous either through their direct application or through society's (inappropriate) reliance on their continued successful operation. Imaginable hazards associated with synthetic biology include; the accidental release of an unintentionally harmful organism or system, the purposeful or accidental design and release of an intentionally harmful organism or system, a future over-reliance on our ability to design and maintain engineered biological systems in an otherwise natural world. In response to these concerns scientists attempt to reduce these risks by; working only with Biosafety Level 1 organisms and components in approved research facilities, working to educate and train a responsible generation of biological engineers and scientists, learning what is possible (at what cost) using simple test systems. We cannot be sure though how this technology will be used maliciously, and since there is really no structure or system with 100% enforcement of ethics in this field, it may only be a matter of time before we experience either an extinction-level event, or all life will permanently be changed due to this technology. Synthetic Biologists believe that the understanding and abilities to be gained from synthetic biology justifies its exploration and development.

So do we inherit and inertly pass along the living world or do we have a responsibility to interact and change it forever in unseen ways at the atomic and molecular level? If we are going to interact with the living world we should at least ground this interaction at a level of resolution that allows for the precise description of our actions and their consequences.

A growing number of artists are attracted to synthetic biology as a technique and also because of the interesting ethical questions it raises. Many of these artists work directly with research scientists. Their creations add a cultural counterbalance to the field's tendency to view life like circuitry, a utilitarian perspective that increasingly drives synthetic biology and, they say, informs the public's understanding of it. They find themselves uniquely placed to ask hard questions about the ethical and social issues raised by synthetic biology. While special interests that want to either promote or condemn the nascent science have been eager to fund artistic interpretations of it, they are finding they may not get the results they hoped for.

The stated goal of synthetic biology is to extend or modify the behavior of organisms and engineer them to perform new tasks. One useful analogy to conceptualize both the goal and methods of synthetic biology is the computer engineering hierarchy. Within the hierarchy, every constituent part is embedded in a more complex system that provides its context. Design of new behavior occurs with the top of the hierarchy in mind but is implemented bottom-up. At the bottom of the hierarchy are DNA, RNA, proteins, and metabolites (including lipids and carbohydrates, amino acids, and nucleotides), analogous to the physical layer of transistors, capacitors, and resistors in computer engineering. The next layer, the device layer, comprises biochemical reactions that regulate the flow of information and manipulate physical processes, equivalent to engineered logic gates that perform computations in a computer. At the module layer, the synthetic biologist uses a diverse library of biological devices to assemble complex pathways that function like integrated circuits. The connection of these modules to each other and their integration into host cells allows the synthetic biologist to extend or modify the behavior of cells in a programmatic fashion. Although independently operating engineered cells can perform tasks of varying complexity, more sophisticated coordinated tasks are possible with populations of communicating cells, much like the case with computer networks.

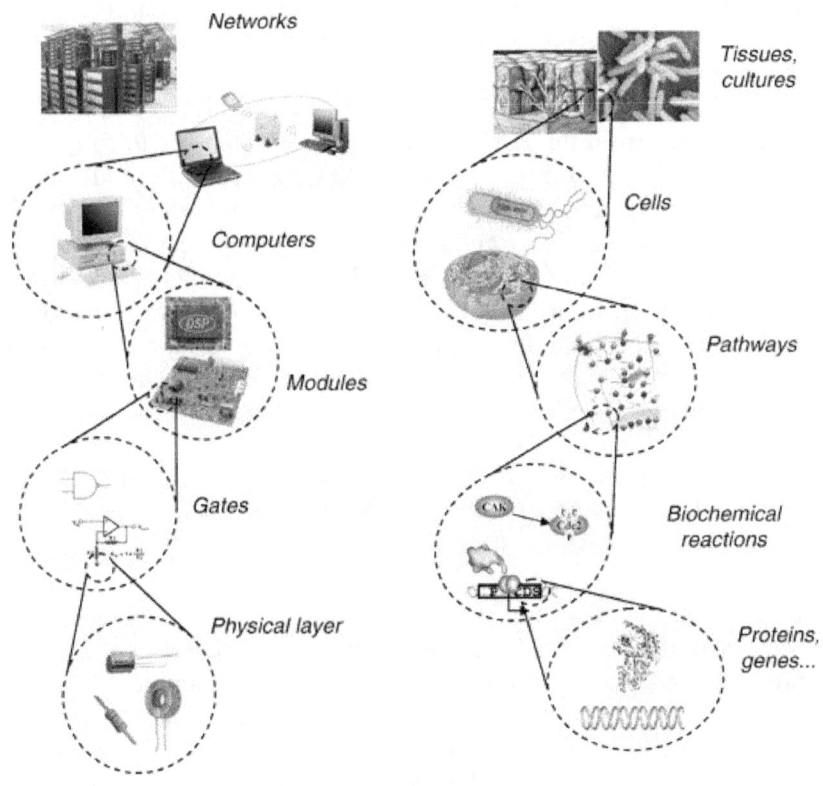

A possible hierarchy for synthetic biology is inspired by computer engineering.

It is useful to apply many existing standards for engineering from well-established fields, including software and electrical engineering, mechanical engineering, and civil engineering, to synthetic biology. Methods and criteria such as standardization, abstraction, modularity, predictability, reliability, and uniformity greatly increase the speed and tractability of design. However, care must be taken in directly adopting accepted methods and criteria to the engineering of biology.

Building biological systems entails a unique set of design problems and solutions. Biological devices and modules are not independent objects, and are not built in the absence of a biological environment. Biological devices and modules typically function within a cellular environment. When synthetic biologists engineer devices or modules, they do so using the resources and machinery of host cells, but in the process also modify the cells themselves.

A major concern in this process is our present inability to fully predict the functions of even simple devices in engineered cells and construct systems that perform complex tasks with precision and reliability. The lack of predictive power stems from several sources of uncertainty, some of which signify the incompleteness of available information about inherent cellular characteristics. The effects of gene expression noise, mutation, cell death, undefined and changing extracellular environments, and interactions with cellular context currently hinder us from engineering single cells with the confidence that we can engineer computers to do specific tasks.

However, most applications or tasks we set to our synthetic biological systems are generally completed by a population of cells, not any single cell.

In a synthetic system, predictability and reliability may be achieved in two ways: statistically by utilizing large numbers of independent cells or by synchronizing individual cells through intercellular communication to make each cell more predictable and reliable. More importantly, intercellular communication can coordinate tasks across heterogeneous cell populations to elicit highly sophisticated behavior. Thus, it may be best to focus on multicellular systems to achieve overall reliability in performing complex tasks.

Living organisms consist of microscopic compartments called cells that usually have largest dimensions in the range of 2-20 micrometers (about 1-10% of the diameter of a human hair). Because cells are so small, ordinary biological samples typically contain enormous numbers of them. An adult human body, for example, contains approximately 100,000,000,000,000 (one hundred trillion) cells, and a liquid culture of E. coli bacteria near the end of logarithmic growth contains several billion cells per milliliter.

Many synthetic biology projects involve engineering the components that carry out such interactions between cells and the outside world. Nevertheless, to manipulate these processes we must also work with the intracellular machinery. Thus cells supply the framework or "chassis" for synthetic biology. At the present time most synthetic biology experiments involve simple single celled organisms like the bacterium Escherichia coli or the yeast Saccharomyces cerevisiae. Many researchers expect the development of drastically modified 4 Cells from eukaryotes ("higher" organisms) range from 10-100 mm, with animal cells normally smaller than plant cells. However, egg cells are notably larger (even visible to the naked eye), muscle cells can fuse into large multinuclear syncytia that extend for several centimeters, and – famously – some nerve cells in giraffes extend for the entire length of the neck (though their diameters are still in the micrometer range). There are also some extraordinarily large unicellular bacteria and protozoa. All these cases represent relatively rare exceptions to the general rule or use of single celled organisms given above. Exceptions are degradative processes catalyzed by secreted enzymes, including complex materials like DNA and viruses.

Molecular biology and biochemistry provide a dynamic picture of life's processes at atomic dimensions. The first steps in moving to this high-resolution view of living organisms were taken with the invention of the light microscope about 350 years ago. Visible-light imaging, however, cannot reveal individual atoms and molecules, even with the most powerful lenses. Thus scientists had to turn to chemistry to develop detailed descriptions of the structures and processes that occur living cells. Elaborate 20th-century instruments such as X-ray diffractometers, electron microscopes and atomic-force microscopes have helped to bridge the gap between what we can see with our eyes, aided by optical microscopes, and what we can deduce from chemical principles. Now, at the beginning of the 21st century, no scale of biological structure – from Angstroms to meters – escapes our scrutiny. That means that scientists can confidently engineer living systems, beginning with an increasingly thorough understanding of the natural prototypes and a complete examination of the systems built.

So far scientists have surveyed the contents and processes of naturally evolved biological systems at the molecular level. Recognizing the important distinction between small molecules and macromolecules leads to an understanding of how complex structures can be built from a relatively small set of low-molecular-weight components (amino acids, nucleotides, sugars and lipid monomers). Intermediary metabolism provides these building blocks as well as energy-rich molecules like ATP that can drive the uphill reactions needed to assemble them into the macromolecules. These in turn perform key biological functions such as catalysis, defense, structure-creation, and information storage. Nucleic acids nearly monopolize this latter task, though epigenetic also participate. Most of the remaining functions can be carried out by proteins, but polysaccharides play a major role in forming structures, and lipids are indispensable for providing the membranes that define cells themselves. In eukaryotes they also create important sub-compartments within the cells. Genetic information, stored in DNA, enables cells to produce specific proteins (via mRNA synthesis) as well as functional RNA molecules. By controlling the ensemble of these "actor" macromolecules, cells manage their own behavior. Synthetic biology aims to modify or extend such behaviors by introducing new genes (and genetic regulatory elements). Ultimately foreseen are reconstructions of cellular genomes that encompass larger and larger numbers of functional genes, perhaps culminating in a totally synthetic cell.

Biological devices

Biologists are familiar with manipulation of genes and proteins to probe their properties and understand biological processes. Synthetic biologists must also manipulate the material elements of the cell, but they do so for the purpose of design, to build synthetic biological systems. Synthetic biologists design complex systems by combining basic design units that represent biological functions. The notion of a device is an concept overlaid on physical processes that allows for decomposition of systems into basic functional parts. Biological devices process inputs to produce outputs by regulating information flow, performing metabolic and biosynthetic functions, and interfacing with other devices and their environments. Biological devices represent sets of one or more biochemical reactions including transcription, translation, protein phosphorylation, allosteric regulation, ligand/receptor binding, and enzymatic reactions. Some devices may include many diverse reactants and products (e.g. a transcriptional device includes a regulated gene, transcription factors, promoter site, and RNA polymerase), or very few (e.g. a protein phosphorylation device includes a kinase and a substrate). The diverse biochemistries underlying the different devices each provide their own advantages and limitations. Particular device types may be more suitable for specific biological activities and timescales. Although the diversity of biochemical reactions makes it difficult to interface devices, it enables the construction of complex systems with rich functionalities.

Scientists can appropriate unaltered devices from nature, for example, the transcription factor(s) and promoter of a regulated gene, for use in our artificial systems, or they can build devices from modified biochemical reactants. The first synthetic biological devices controlled transcription by modifying promoter sequences to bind novel transcriptional activators and repressors in prokaryotes and eukaryotes (Baron et al, 1997; Lutz and Bujard, 1997; Becskei and Serrano, 2000; Lutz et al, 2001). Recent efforts also focused on non-transcriptional control. Non-coding RNAs can activate or silence gene expression by regulating translation events in prokaryotes (Isaacs et al, 2004). Synthetic biologists also extended the use of non-coding regulatory RNAs to eukaryotes (Bayer and Smolke, 2005). The non-coding RNAs introduced into *Saccharomyces cerevisiae* consisted of aptamer domains that bind specific ligands and antisense domains that target mRNAs. The non-coding RNA undergoes a conformational change upon binding its ligand that enables the antisense domain to bind the target mRNA, thus modulating translation.

Devices that control transcription and/or translation to produce an output are relatively flexible and easy to build, as nucleotide sequences directly determine the specificity and efficiency of interactions. The ability to concatenate and edit nucleotide sequences at will facilitates the connection of essentially any two devices (as when the promoter in one device is placed in front of the gene encoding a repressor or activator in another device). Transcriptional control systems have many additional useful properties, including signal amplification, combinatorial control by multiple transcription factors, and control of multiple downstream targets. However, changes in output are relatively slow because the process of gene expression occurs on a timescale of minutes owing to the large number of biochemical reactions required to synthesize even a single protein. Achieving a detectable change in output requires many of these protein synthesis events, so these devices consume a large amount of cellular resources during their function.

Devices derived from protein–ligand and protein–protein interactions have different input and output characteristics than transcriptional control devices, and their construction may require more involved modification of their natural substrates. An early approach to building devices based on protein interactions was modular recombination to create allosteric gating mechanisms (Dueber *et al*, 2003). Other researchers created protein switches by inserting allosteric domains into existing enzymes (Guntas and Ostermeier, 2004). Random domain insertion followed by directed evolution produced hybrid enzymes (combining maltose-binding protein with beta lactamase) in which maltose altered lactamase catalytic activity by as much as 600-fold (Guntas *et al*, 2005).

Using mutagenesis guided by computational chemistry, protein engineers rationally redesigned and constructed altered periplasmic *Escherichia coli* receptors that respond to a variety of extracellular ligands like TNT, L-lactate, or serotonin instead of their natural ligands (Looger *et al*, 2003), and also converted ribose-binding protein, a sensor protein that lacks enzymatic activity, into a protein that exhibits triose phosphate isomerase activity (Dwyer *et al*, 2004). Construction of these devices utilized computational chemistry to model binding sites and active sites with key residues mutated to confer altered activity (Dwyer and Hellinga, 2004). A geometric search algorithm examines the 3D structure of a protein to locate positions on the polypeptide backbone where placement of prespecified mutated amino-acid side chains simultaneously satisfies all desired geometrical constraints for binding. Designs can be improved iteratively by successive rounds of mutation and geometric search, or by implementing another algorithm that optimizes stereochemical packing of amino-acid side chains. Although these algorithms greatly narrow down the list of requisite mutations, the process must be followed by directed evolution to refine the effectiveness of binding or enzymatic activity. Redesigned receptors can be the initial input devices of extended protein phosphorylation cascades (Looger *et al*, 2003).

In constructing protein interaction devices (Giesecke *et al*, 2006), the proteins must be well characterized to determine where changes, deletions, and replacements of domains can occur, as their 3D structure plays a large role in the nature of their interactions. Connecting protein interaction devices is not a trivial task. Binding interfaces between proteins from different devices must be well matched and one must validate that transfer of information occurs between devices (i.e. a conformational change occurs upon binding or phosphorylation). However, protein interaction devices offer significant design benefits. Binding and enzymatic reactions occur on the sub-second timescale, so changes in output are very fast. It is possible to amplify signals when using protein interaction devices, owing to the reusability of kinases and other enzymes. A minimal amount of protein synthesis can yield devices composed of proteins that undergo repeated interaction events with multiple partners, thus taking up only modest amounts of cellular resources during their function. Additionally, the degree of insulation of protein interaction devices from endogenous cellular processes depends mostly on binding specificity because these devices do not require the protein synthesis machinery of the host cell to produce an output.

Cellular functions of a device are conditioned by the substrate and biochemical reactions chosen. For example, transcriptional and translational devices are easy to connect and are capable of great logical complexity, but such devices cannot be assembled into systems that respond in seconds. Protein interaction devices, however, can provide fast responses. Synthetic biologists will have to combine different types of devices to design the most efficient modules, so future research must establish effective interfacing methods.

Interfacing devices to build synthetic biological modules

A module is a compartmentalized set of devices with interconnected functions that performs complex tasks. In the cell, modules are specific pathways, such as a metabolic pathway or a signal transduction pathway. Ultimately understanding how the function of a module or an entire biological system can be derived from the function of its component parts is very important. Such knowledge will help to establish the biological *rules of composition* to build modules from devices. The rules of composition help determine which device combinations yield the desired logic functions and, more importantly, how to match cellular or physical functions of devices.

Most devices are derived from naturally occurring systems. The difficulty in constructing modules from diverse wild-type devices is that evolution has already optimized them to perform within their natural contexts, so they may not function when connected to each other in an artificial context. Synthetic biologists typically need to change device characteristics (i.e. device physics) in order to produce the desired logical functions when these elements are interfaced. Rational redesign based on mathematical modeling and directed evolution of devices can help match them so they function properly together.

Rational mutations of devices are particularly useful for changing the overall behavior of the system when the properties of these devices are fairly well known. Modifying the kinetics of transcription and translation, operator binding affinity, and binding cooperativity of transcription factors can help generate devices that enable a module to meet desired criteria, such as having a digital step-like response that yields robustness to extrinsic noise (Baron *et al*, 1997; Weiss and Basu, 2002). Successful module function thus often requires alteration of wild-type devices to properly interface them.

Even simple modules can take significant amounts of time and resources to construct from devices, often requiring multiple revisions to optimize behavior. Modeling greatly aids in overcoming module design problems (McAdams and Arkin, 1998; Alon, 2003; Kaern *et al*, 2003). The requisite computational tools use abstractions of biochemical reactions to model devices and typically require the rate constants of those reactions. Direct determination of rate constants *in vivo* is still inaccurate and far from trivial, but this barrier may be overcome by the technique of parameter estimation in biological networks (Ronen *et al*, 2002; Braun *et al*, 2005). Adding or removing devices will change the module and change any previously estimated parameters. Thus, parameters derived in one context may not apply in another. Mechanisms for system redesign and optimization that do not rely on precise parameters must be used.

For system design with incomplete knowledge, it is important to determine which reactions most significantly affect system output. When the system does not function as expected, these reactions are likely to be the best targets for modification. Because the output is typically sensitive to only a few parameters, the set of candidate reactants for mutagenesis can be narrowed to a manageable size. Sensitivity analysis achieves these aims, as demonstrated in a recent study with synthetic gene networks (Feng *et al*, 2004). Sensitivity analysis does not derive actual rate constants, but determines which ones are important.

Examples of synthetic biological modules

Modules span a continuum of complexity. A simple module may be an instance of a recurring network motif. Complex modules may be composed of many motifs and do not necessarily occur frequently. The variety of motifs allow for many modes of regulation in natural systems (Milo *et al*, 2002, 2004). It may be tempting to use naturally occurring network motifs and their modules for introducing new behavior into organisms. However, natural modules are not optimized for operation within a cellular context that is not their own, and in fact may not be functional at all. In addition, they are difficult to modify and it may be impossible to find an appropriate natural module that performs a desired task. Synthetic modules can be more readily understood in a quantitative fashion than natural modules. They have well-defined boundaries and points of connection to other modules, and are therefore amenable to insulation from most cellular processes. They can thus be more easily removed, replaced, or altered than natural modules. Most importantly, as we are not limited by natural examples, simple synthetic modules are extensible to functions beyond those available in nature. In this section, we discuss the major prototype synthetic modules: transcriptional regulation networks, protein signaling pathways, and metabolic networks.

Tremendous progress has been made in developing prototype modules with non-trivial behaviors. Modules with certain motifs have been shown to function as desired (e.g. ultrasensitive cascades, feedforward motifs, and bistable switches). These motifs have been engineered to operate correctly using the same tools applied to connecting devices, as described in the previous section. Much work is still needed to engineer certain other motifs to function reliably and predictably, for example, oscillators. Future research will focus on the integration of these basic motifs to form complex modules with interesting high-level functions (counters, adders, multiple signal integrators). Proper interfacing of diverse module types (transcription regulation, protein signaling, and metabolic networks) can extend their function and such procedures will make reliable, definable connections to cellular context and thus generate well-designed cells.

Engineering cells: interfacing with cellular context and connecting modules

The functional behavior of a module in a cell depends not only on its component devices and their connectivity (wiring), but also on the cellular context in which the module operates. Relevant cellular context includes general biochemical processes such as DNA and RNA metabolism, availability of amino acids, ATP levels, protein synthesis, cell cycle and division, and specific processes such as endogenous signaling pathways that may interact with devices in the exogenous module. As a consequence, the same gene circuit can have different behavior in slightly different host cell strains (Endy, 2005). In addition, integration and function of a module in a host cell may fundamentally affect host cell processes, thus altering the cellular context, which may then recursively alter the behavior of the module. The situation is further complicated by the integration and function of additional modules.

Because synthetic modules and endogenous cellular processes condition each other's behavior, any fluctuations in the host cell processes are relayed to the module and affect its output and vice versa. This presents a problem for engineering predictable, reliable biological systems. One approach to solving this problem is to take the notion of modularity to heart. Modularity is used in other engineering disciplines to insulate interacting systems from each other and render them interchangeable. According to this notion, inserted modules would function best if the number of interactions between the module and the host cell are minimized. Any remaining interactions should ideally be very predictable. Specifying and standardizing those remaining interactions can ensure the portability of the modules, and allow them to be engineered independently of host cells. Using this approach, module function would ideally become independent of cellular context. The host cell would only act to process resources and protect the module from the extracellular environment.

If achievable, insulation of modules is useful, but must be tempered with a drive to understand and take into account a module's connection to the host's cellular context. Although simplification, specification, and standardization make engineering easier, it may not be advantageous to hide all the information about the host cell. Conceptualizing the operation of a module as completely, or nearly completely, disconnected from cellular context cannot sufficiently define module function. Part of what defines living systems is the integration of their parts. Engineering any part of an organism must at some level take the entire organism into account. Thus, for modular composition to work, we need to have abstractions that incorporate the notion of cellular context into the definition of a module's function. We must have a quantifiable way to encode context dependence for a given module and functionally compose the context dependence of multiple modules.

Rational redesign, directed evolution, and modeling, which played such important roles in assembling devices into modules, can play similar roles in interfacing modules with the cellular context of a host cell. Combining parameter estimation techniques with metabolic flux balance analysis to take into account relevant contextual features may be a promising path. It may be necessary to quantify the effects of context iteratively after the addition of each module to glean a more accurate description of cells that harbor our modules of interest. Insertion of some modules may be easy, but others will be difficult to insert and will require a great deal of modification of the host cell for optimal compatibility. Adding multiple modules piecemeal in this fashion could be prohibitively difficult. The parallel with software engineering is instructive: this is like adding foreign software to an operating system (OS), each new program with its own patch. Adding too many programs often leads to system-wide instability. A better strategy in this case would be to build a new OS. For synthetic biology, this means engineering a new organism by synthesizing its genome.

Synthetic genomes

Advances in DNA synthesis will drive progress in the construction of synthetic genomes to provide a reliable method for building an entirely artificial organism (Zimmer, 2003). Constructing an organism wholesale has certain advantages. We may choose to include only the functions and pathways that we want, either for simplicity, to cut down on evolutionary baggage, or to make the genome easier to customize. The system would allow users to plug in any desired module, and implement hooks for extensibility of features and perhaps dependence on laboratory supplements in the media for safety. A synthetic genome, like a computer's OS, needs particular conditions and substrates to operate, but may incorporate only desired features. Just as with operating systems, synthetic genomes would be delivered with manuals and full documentation. If there is no simple migration path from one genome or OS to another, then it would be simpler to change the OS, or synthesize a novel genome than to try and convert multiple applications or modules. A good place to start is with the simplest known genomes, that is, viruses. Synthetic virus genomes have been constructed *de novo* and successfully tested for the ability to generate infectious viruses. Poliovirus cDNA was synthesized, transcribed into viral RNA, replicated in cell extracts, and injected into transgenic mice (Cello et al, 2002). The synthetic poliovirus induced the same neurovirulent phenotype as wild-type poliovirus. The ΦX174 bacteriophage genome (5386 bp) has also been synthesized, and used to infect *E. coli* cells (Smith et al, 2003). In other work, the T7 bacteriophage genome has been refactored to make the virus simpler to model and more amenable to manipulation (Chan et al, 2005).

Beyond viruses, progress has been made towards identifying and removing non-essential genes in the genomes of *E. coli* as well as *Mycoplasma genitalium*, which has one of the smallest known genomes (Pennisi, 2005). The resulting minimal genomes may yield organisms that serve as ideal vessels for synthetic gene networks. Genome synthesis will make it possible to fabricate minimal cells, creating the simplest possible contexts for inserting new modules. This leads to the question: can man design a better 'cell chassis'? We still need to determine if the properties of a minimal cell chassis are those of an optimal cell chassis. Simplification of its genome may not yield a generic organism that operates with precision and reliability.

Synthetic biologists compartmentalized portions of a pulse-generating network into separate sender and receiver cell populations where sender cells produce the signal that triggers a pulse in the receiver cells. Sender cells were placed in a fixed position with receiver cells surrounding them on solid-phase media. As the pulse-generating network can sense different rates of increase of the signal, receiver cells near the sender cells produce a detectable pulse response, whereas cells farther away did not respond as well. This signal processing capability is useful for designing systems that require communication mechanisms with fine-tuned localized responses to a diffusible signal. Researchers also demonstrated programmed pattern formation using a band-detect network (Basu et al, 2005). The network integrates an artificial cell–cell communication system with multiple transcriptional genetic inverters that have dissimilar repression efficiency in a feedforward motif. The system only responds to a predefined range of signal concentration, and this range is engineered by altering different parameters of the network. By placing sender cells in predetermined positions with variants of undifferentiated band-detect cells surrounding them on solid-phase media, the band-detect system generated elaborate spatial patterns in the form of hearts, clovers, bullseyes, and ellipses. More complex multicellular patterns with enhanced robustness will be achieved by integrating multiple communication signals and feedback mechanisms into the system.

Communication between many different types or populations of cells requires not only insulation between multiple communication channels, but also interoperability. Building such complexity requires the use of several types of signaling components, more kinds of host cells, multiway communication, and improvements in precision and reliability. Designing communication systems in a multicellular environment entails balancing the sensitivities of the intercellular elements and reducing the crosstalk between those signals. Engineered yeast cells synthesize the plant hormone cytokinin in a positive feedback loop to achieve quorum sensing. Cytokinin diffuses into the environment and activates a hybrid exogenous/endogenous phosphorylation network that triggers production of more cytokinin, leading to population-dependent gene expression.

The design, construction, and testing of multicellular systems will ultimately have a significant impact on central problems both in biology and computing. Synthetic biology provides unique opportunities and powerful tools to investigate spatiotemporal patterning of gene expression and the mechanisms governing coordinated cell behavior during development of multicellular organisms. We will eventually be able to build synthetic model systems of biological development. Synthetic biology will provide useful and manipulatable model systems to answer a central question in computing and biology: how can complex and robust global behavior emerge from the interactions of large numbers of unreliable, locally communicating components?

Synthetic biology distinguishes itself from other engineering and scientific disciplines in both its approach and its choice of object. This emerging field uses the insights of scientific biological inquiry but formulates new rules for engineering purposes. Synthetic biology should be considered a hybrid discipline, combining elements of both engineering and science to achieve its goal of engineering synthetic organisms.

Living systems are highly complex, and we currently lack a great deal of information about how these systems work. One reason is that biological systems possess a degree of integration of their parts far greater than that of non-living systems. Breaking down organisms into a hierarchy of composable parts, although useful as a tool for conceptualization, should not lull the reader into thinking that these parts can be assembled *ex nihilo*. Because we do not yet know how to confer the properties of life onto an aggregate of physically dynamic, but 'dead' material systems, composing artificial living systems requires the use and modification of natural ones. Therefore, assembly of parts occurs in a biological setting, within an existing cellular context. This has profound implications for the abstraction of biological components into devices and modules and their use in design.

In computer engineering, it is possible to isolate hardware design (computer architecture) from software design (programming), making it easy to implement different behaviors on the same physical platform. The software analogy used earlier was instructive for understanding the development of synthetic genomes and the role of modules within that framework. However, at this stage of synthetic biology, 'programming' actually means altering the hardware itself. Reprogramming a cell involves the creation of synthetic biological components by adding, removing, or changing genes and proteins. Direct interaction with the hardware of engineered cells thus more closely resembles the incorporation of a user-designed application-specific integrated circuit (ASIC) into a computer. With ASICs, users can easily extend computer function by designing specialized hardware logic that is then fabricated on a custom silicon chip. In contrast to writing computer software, extending computer hardware by incorporating ASICs may require modifications to existing hardware (e.g. adding an extra power supply or enlarging the computer chassis). Similarly, the addition of synthetic gene networks (new cellular hardware) to host cells requires careful attention to cellular context (existing cellular hardware).

In the design, fabrication, integration, and testing of new cellular hardware, synthetic biologists must use tools and methods derived from experimental biology. However, experimental biology still has not progressed to the point where it can provide an unshakable foundation for synthetic biology the way solid-state physics did for electrical engineering. As a result, design of synthetic biological systems has become an iterative process of modeling, construction, and experimental testing that continues until a system achieves the desired behavior. The process begins with the abstract design of devices, modules, or organisms, and is often guided by mathematical models. The synthetic biologist then tests the newly constructed systems experimentally. However, such initial attempts rarely yield fully functional implementations because of incomplete biological information. Rational redesign based on mathematical models improves system behavior in such situations. Directed evolution is a complimentary approach, which can yield novel and unexpected beneficial changes to the system. These retooled systems are once again tested experimentally and the process is repeated as needed. Many synthetic biological systems have been engineered successfully in this fashion because the methodology is highly tolerant to uncertainty. Synthetic biology will benefit from further such development and the creation of new methods that manage uncertainty and complexity.

Standardization involves establishing definitions of biological functions and methods for identifying biological parts, as with the registry of standard biological parts (Knight, 2002). Efforts in systems biology targeted at classification and categorization of genome elements can help define biological functions. Decoupling involves the decomposition of complicated problems into simpler problems, implemented by breaking down complex systems into its simpler constituents, and separating design from fabrication. Abstraction includes establishing hierarchies of devices and modules that allow separation and limited exchange of information between levels, and developing redesigned and simplified devices and modules, as well as libraries of parts with compatible interfaces.

It may not be possible or desirable to fully separate and insulate biological devices and modules from each other and from the machinery of the host cells. The notions of standardization, decoupling, and abstraction must therefore be recast to better reflect the complexity of the cellular context. We will require not only standards for specific biological functions, but also standards for the states of contextual cellular elements (e.g. ATP used when a cell divides). Although decoupling design from fabrication is useful, breaking down complex systems into many simpler ones will miss the connections between the simple elements and the radical interconnectedness of cellular context with each inserted module. Accordingly, abstractions of device and module function must include cellular context.

A biological device has no meaning isolated from a module; a module has no meaning isolated from a cell; a cell has no meaning isolated from a population of cells. This contextual dependence is an essential feature of living systems and is not an impasse, but rather a bridge to the successful engineering of living systems. As with the uncertainty principle in quantum mechanics, it may be prudent to treat some biological uncertainties as fundamental properties of individual cell behavior (e.g. gene expression noise, context dependence, fluctuating environments).

The fact that scientists always use populations of synthetic cells to complete tasks means that the criteria of reliability and predictability should apply at the cell population level. As long as a significant number of the cell population performs the desired task, the unpredictability of events occurring at the molecular level should have minimal effect. Design and fabrication methods that take into account uncertainty and context dependence will likely lead to on-demand, just-in-time customization of biological devices and components, which need not behave perfectly. Building imperfect systems is generally thought of as acceptable, as long as they perform tasks adequately. Synthetic biology attempts to use the strategies that make biological systems versatile and robust as part of its own design practices. The success of synthetic biology will depend on its capacity to surpass traditional engineering, blending the best features of natural systems with artificial designs that are extensible, comprehensible, user-friendly, and most importantly implement stated specifications to fulfill user goals.

8 DNA, GNA, PNA

For all the magnificent diversity of life on this planet, ranging from tiny bacteria to majestic blue whales, from sunshine-harvesting plants to mineral-digesting endoliths miles underground, only one kind of *"life as we know it"* exists. All these organisms are based on nucleic acids—DNA and RNA—and proteins, working together more or less as described by the so-called central dogma of molecular biology: DNA stores information that is transcribed into RNA, which then serves as a template for producing a protein. The proteins, in turn, serve as important structural elements in tissues and, as enzymes, are the cell's workhorses.

Yet scientists dream of synthesizing life that is utterly alien to this world—both to better understand the minimum components required for life (as part of the quest to uncover the essence of life and how life originated on earth) and, frankly, to see if they can do it. That is, they hope to put together a novel combination of molecules that can self-organize, metabolize (make use of an energy source), grow, reproduce and evolve.

Yes, just what we need: a synthetic 'life' that could evolve (or be turned into) a bioweapon more powerful than any from or modified from nature.

Nucleic acids are biological molecules essential for life, and include DNA (deoxyribonucleic acid) and RNA (ribonucleic acid). Together with proteins, nucleic acids make up the most important macromolecules; each is found in abundance in all living things, where they function in encoding, transmitting and expressing genetic information.

Nucleic acids were discovered by Friedrich Miescher in 1869. Experimental studies of nucleic acids constitute a major part of modern biological and medical research, and form a foundation for genome and forensic science, as well as the biotechnology and pharmaceutical industries.

Deoxyribonucleic acid (**DNA**) is a nucleic acid containing the genetic instructions used in the development and functioning of all known living organisms (with the exception of RNA viruses). The DNA segments carrying this genetic information are called genes. Likewise, other DNA sequences have structural purposes, or are involved in regulating the use of this genetic information. Along with RNA and proteins, DNA is one of the three major macromolecules that are essential for all known forms of life.

DNA consists of two long polymers of simple units called nucleotides, with backbones made of sugars and phosphate groups joined by ester bonds. These two strands run in opposite directions to each other and are therefore anti-parallel. Attached to each sugar is one of four types of molecules called nucleobases (informally, *bases*). It is the sequence of these four nucleobases along the backbone that encodes information. This information is read using the genetic code, which specifies the sequence of the amino acids within proteins. The code is read by copying stretches of DNA into the related nucleic acid RNA in a process called transcription.

Within cells, DNA is organized into long structures called chromosomes. During cell division these chromosomes are duplicated in the process of DNA replication, providing each cell its own complete set of chromosomes. Eukaryotic organisms (animals, plants, fungi, and protists) store most of their DNA inside the cell nucleus and some of their DNA in organelles, such as mitochondria or chloroplasts. In contrast, prokaryotes (bacteria and archaea) store their DNA only in the cytoplasm. Within the chromosomes, chromatin proteins such as histones compact and organize DNA. These compact structures guide the interactions between DNA and other proteins, helping control which parts of the DNA are transcribed.

Bioinformatics involves the manipulation, searching, and data mining of biological data, and this includes DNA sequence data. The development of techniques to store and search DNA sequences have led to widely applied advances in computer science, especially string searching algorithms, machine learning and database theory. String searching or matching algorithms, which find an occurrence of a sequence of letters inside a larger sequence of letters, were developed to search for specific sequences of nucleotides. The DNA sequenced may be aligned with other DNA sequences to identify homologous sequences and locate the specific mutations that make them distinct.

These techniques, especially multiple sequence alignment, are used in studying phylogenetic relationships and protein function. Data sets representing entire genomes' worth of DNA sequences, such as those produced by the Human Genome Project, are difficult to use without the annotations that identify the locations of genes and regulatory elements on each chromosome.

Regions of DNA sequence that have the characteristic patterns associated with protein- or RNA-coding genes can be identified by gene finding algorithms, which allow researchers to predict the presence of particular gene products and their possible functions in an organism even before they have been isolated experimentally. Entire genomes may also be compared which can shed light on the evolutionary history of particular organism and permit the examination of complex evolutionary events.

Types of Nucleic Acids

There are two main types of nucleic acid: deoxyribonucleic acid (DNA) and ribonucleic acid (RNA). They differ by number of strands, by amino acids, by sugars and by use.

RNA serves as a messenger between DNA and the mechanics of protein-building, allowing DNA to stay safely in the nucleus. DNA differs from RNA in that DNA is double-stranded instead of single-stranded; additionally, two of DNA's four bases (nucleotides) are different, as is its sugar component.

There are two types of RNA: messenger (mRNA), which transfers genetic information from the nuclear DNA to the ribosome: and transfer (tRNA), which attaches an individual amino acid to the mRNA.

Artificial Nucleic Acids

Artificial nucleic acids differ from naturally occurring DNA and RNA by the types of receptors onto which the amino acids attach. Names of such nucleic acids include peptide nucleic acid (PNA), glycol nucleic acid (GNA) and threose nucleic acid (TNA).

Bacterial DNA

Bacterial genetic code is usually carried in a single chromosome of double-stranded, circular DNA (as opposed to plants' and animals' linear DNA).

Circular DNA

The difference between linear and circular DNA (in which the DNA hoop circles and closes on itself) lies not in the DNA's appearance but in circular DNA's greater ability to replicate.

Virus Genetic Code

Viruses carry either DNA or RNA, either of which can be single-stranded or double-stranded. Hepadnaviridae, a group of viruses which can cause liver problems, carries both single- and double-stranded DNA.

TNA

TNA is just one of many nucleic acids that may have been important in the first life on Earth. Below are three others.

PNA

PNA (peptide nucleic acid) a synthetic hybrid of protein and DNA, could form the basis of a new class of drugs—and of artificial life unlike anything found in nature. PNA ditches the sugar in its backbone and inserts a peptide instead, so it is more closely related to proteins. Like DNA, it can form double strands with itself, as well as with DNA or RNA, making it a promising genetic system (Science, DOI:10.1126/science.1174577). Several classes of nucleic acid analogs have been reported, but no synthetic informational polymer has yet proven responsive to selection pressures under enzyme-free conditions. Here, is introduced an oligomer family that efficiently self-assembles by means of reversible covalent anchoring of nucleobase recognition units onto simple oligo-dipeptide backbones [thioester peptide nucleic acids (tPNAs)] and undergoes dynamic sequence modification in response to changing templates in solution. The oligomers specifically self-pair with complementary tPNA strands and cross-pair with RNA and DNA in Watson-Crick fashion. Thus, tPNA combines base-pairing interactions with the side-chain functionalities of typical peptides and proteins. These characteristics might prove advantageous for the design or selection of catalytic constructs or biomaterials that are capable of dynamic sequence repair and adaptation. It is also easy to make long PNA molecules in the conditions of prebiotic Earth, even at temperatures of 100 °C (*Proceedings of the National Academy of Sciences*, DOI: 10.1073/pnas.97.8.3868).

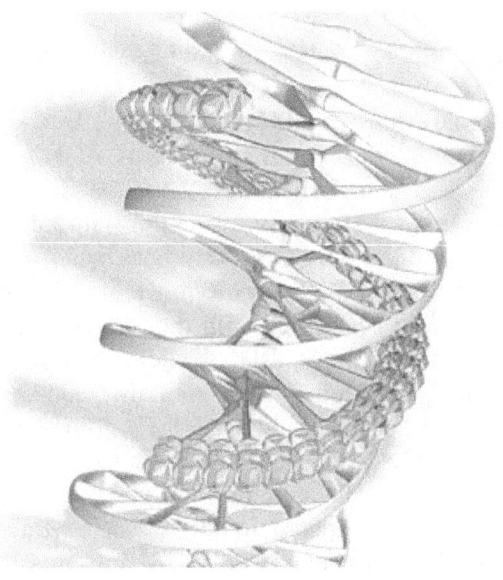

Peptide nucleic acid (gold) readily enters DNA's major groove to form triple-stranded and other structures with DNA, allowing it to modify the activity of genes in new ways. Image: Jean-Francois Podevin

GNA

GNA (glycol nucleic acid) is even simpler than TNA, with just three carbon atoms in its backbone, yet can still form helical molecules, much like DNA. GNA is not known to occur naturally; it is synthesized chemically.

DNA and RNA have a deoxyribose and ribose sugar backbone, respectively, whereas GNA's backbone is composed of repeating glycol units linked by phosphodiester bonds. The glycol unit has just three carbon atoms and still shows Watson-Crick base pairing. Interestingly, the Watson-Crick base pairing is much more stable in GNA than its natural counterparts DNA and RNA as it requires a high temperature to melt a duplex of GNA.

It is possibly the simplest of the nucleic acids, so making it a hypothetical precursor to RNA.

Glycol nucleic acid (GNA), with a nucleotide backbone comprising of just three carbons and the stereocenter derived from propylene glycol (1,2-propanediol), is a structural analog of nucleic acids with intriguing biophysical properties, such as formation of highly stable antiparallel duplexes with high Watson–Crick base pairing fidelity.

Previous crystallographic studies of double stranded GNA (dsGNA) indicated two forms of backbone conformations, an elongated M-type (containing metallo-base pairs) and the condensed N-type (containing brominated base pairs). A new crystal structure of a GNA duplex at 1.8 Å resolution from self-complementary $3'$-CTC^{Br}UAGAG-$2'$ GNA oligonucleotides reveals an N-type conformation with alternating gauche–anti torsions along its ($O3'$–$C3'$–$C2'$–$O2'$) backbone. To elucidate the conformational state of dsGNA in solution, molecular dynamic simulations over a period of 20ns were performed with the now available repertoire of structural information. Interestingly, dsGNA adopts conformational states in solution intermediate between experimentally observed backbone conformations: simulated dsGNA shows the all-gauche conformation characteristic of M-type GNA with the higher helical twist common to N-type GNA structures. The so far counterintuitive, smaller loss of entropy upon duplex formation as compared to DNA can be traced back to the conformational flexibility inherent to dsGNA but missing in dsDNA. Besides extensive interstrand base stacking and conformational preorganization of single strands, this flexibility contributes to the extraordinary thermal stability of GNA.

ANA

ANA (amyloid nucleic acid) consists of nucleic acids attached to amyloid proteins, infamous for their role in Alzheimer's disease. ANA fibers have been suggested as the first organisms, because the amyloid could protect genetic material contained within. Nucleic acids promote amyloid formation in diseases including Alzheimer's and Creutzfeldt-Jakob disease. However, it remains unclear whether the close interactions between amyloid and nucleic acid allow nucleic acid secondary structure to play a role in modulating amyloid structure and function. It is common to have used a simplified system of short basic peptides with alternating hydrophobic and hydrophilic amino acid residues to study nucleic acid - amyloid interactions. Employing biophysical techniques including X-ray fiber diffraction, circular dichroism spectroscopy and electron microscopy we show that the polymerized charges of nucleic acids concentrate and enhance the formation of amyloid from short basic peptides, many of which would not otherwise form fibers. In turn, the amyloid component binds nucleic acids and promotes their hybridization at concentrations below their solution K_d, as shown by time-resolved FRET studies. The self-reinforcing interactions between peptides and nucleic acids lead to the formation of amyloid nucleic acid (ANA) fibers whose properties are distinct from their component polymers. In addition to their importance in disease and potential in engineering, ANA fibers formed from prebiotically-produced peptides and nucleic acids are thought by some to have played a role in early evolution, constituting the first entities subject to Darwinian evolution.

Microarrays and Nanoarrays

With only a few exceptions, every cell of the body contains a full set of chromosomes and identical genes. Only a fraction of these genes are turned on, however, and it is the subset that is **"expressed"** that confers unique properties to each cell type. **"Gene expression"** is the term used to describe the transcription of the information contained within the **DNA**, the repository of genetic information, into messenger RNA (mRNA) molecules that are then translated into the proteins that perform most of the critical functions of cells. Scientists study the kinds and amounts of mRNA produced by a cell to learn which genes are expressed, which in turn provides insights into how the cell responds to its changing needs. Gene expression is a highly complex and tightly regulated process that allows a cell to respond dynamically both to environmental stimuli and to its own changing needs. This mechanism acts as both an **"on/off" switch** to control which genes are expressed in a cell as well as a **"volume control"** that increases or decreases the level of expression of particular genes as necessary.

Microarrays are a significant advance both because they may **contain a very large number of genes** and because of their **small size**. Microarrays are therefore useful when one wants to survey a large number of genes quickly or when the sample to be studied is small. Microarrays may be used to assay gene expression within a single sample or to compare gene expression in two different cell types or tissue samples, such as in healthy and diseased tissue. Because a microarray can be used to examine the expression of hundreds or thousands of genes at once, it promises to revolutionize the way scientists examine gene expression. This technology is still considered to be in its infancy; therefore, many initial studies using microarrays have represented simple surveys of gene expression profiles in a variety of cell types. Nevertheless, these studies represent an important and necessary first step in our understanding and cataloging of the human genome.

As more information accumulates, scientists will be able to use microarrays and nanoarrays to ask increasingly complex questions and perform more intricate experiments. With new advances, researchers will be able to infer probable functions of new genes based on similarities in expression patterns with those of known genes. Ultimately, these studies promise to expand the size of existing gene families, reveal new patterns of coordinated gene expression across gene families, and uncover entirely new categories of genes. Furthermore, because the product of any one gene usually interacts with those of many others, our understanding of how these genes coordinate will become clearer through such analyses, and precise knowledge of these inter-relationships will emerge. The use of microarrays and nanoarrays may also speed the identification of genes involved in the development of various diseases by enabling scientists to examine a much larger number of genes. This technology will also aid the examination of the integration of gene expression and function at the cellular level, revealing how multiple gene products work together to produce physical and chemical responses to both static and changing cellular needs.

GNA hybridization

Glycerol nucleic acid (GNA) is an interesting alternative base-pairing system based on an acyclic, glycerol-phosphate backbone repeat unit. The question of whether DNA polymerases can catalyze efficient template-dependent synthesis using GNA as the template is of particular interest because GNA is unable to form a stable duplex with DNA. A variety of DNA polymerases were screened for GNA-dependent DNA synthesis. It was found that Bst DNA polymerase can catalyze full-length DNA synthesis on a dodecamer GNA template. The efficiency of DNA synthesis is increased by replacing adenine with diaminopurine in both the GNA template and the DNA monomers and by the presence of manganese ions. We suggest that the BstDNA polymerase maintains a short, transient region of base-pairing between the DNA product strand and the GNA template, but that stable duplex formation between product and template strands is not required for template-dependent polymerization.

Nucleic acid analogs with altered backbones or bases are of significant interest in the search for biopolymers with novel chemical and biological properties, and many such analogs have been designed and synthesized.

Evaluation of the hybridization and nuclease-resistance properties of these synthetic nucleic acids has led to several nucleic acid analogues with potential biological applications. However, much less is known about the potential for information transfer between synthetic nucleic acid systems and the modern DNA/RNA system via template-dependent polymerization, a crucial aspect of the chemical etiology of nucleic acids and directed evolution of functional biopolymers based on a synthetic nucleic acid system.

Recently, scientists have studied the enzymatic synthesis of $(3' \rightarrow 2')$ α-l-threose nucleic acid (TNA). TNA was discovered by Eschenmoser and coworkers during a systematic evaluation of the base-pairing properties of nucleic acids containing alternative sugar-phosphate backbones. TNA was found to be capable of forming stable, antiparallel duplexes with RNA, DNA, and itself, a surprising property for a nucleic acid analog with a shorter backbone repeat unit than DNA or RNA. This remarkable intersystem duplex formation has been considered to be the basis of possible information transfer between TNA and DNA/RNA. Later studies revealed that α-l-threose nucleoside 3'-triphosphates were substrates for efficient template-dependent enzymatic polymerization by Therminator DNA polymerase, a mutated archaeal family B DNA polymerase. Therminator DNA polymerase was also shown to be an efficient and accurate TNA-dependent DNA polymerase. The information transfer between TNA and DNA catalyzed by polymerases provides support for a possible role of TNA as a progenitor of DNA/RNA.

The search for nucleic acid analogs with even simpler backbones led to studies of the glycerol nucleic acids (GNAs), which have a three-carbon, acyclic backbone. The S isomer of GNA can form stable antiparallel duplexes with itself and RNA, but not with DNA of the sequence 3'-taa aat tta tat tat taa-2' (lowercase type denotes the GNA sequence). These properties suggested that direct sequence information exchange between GNA and DNA might not be possible. In the same study, a panel of polymerases was screened for their ability to catalyze DNA synthesis on GNA templates. Surprisingly, even without stable duplex formation between GNA and DNA, as demonstrated by thermal denaturation studies using optical hyperchromicity and CD spectroscopy, Bst DNA polymerase is able to catalyze DNA synthesis on a GNA template with good fidelity.

These results demonstrate template-dependent synthesis in the absence of a stable product–template duplex.

It is generally assumed that stable duplex formation between the template and the growing product strand is required to allow enzymatic DNA polymerization to proceed. In contrast, we have found that Bst polymerase, a Family A DNA polymerase produced by thermophilic BACILLUS STEAROTHERMOPHILUS, is able to carry out template-dependent DNA synthesis on a GNA template, which does not form a stable duplex with DNA. The catalytic efficiency of Bst polymerase on a GNA template increases in the presence of Mn^{2+} ions, which is known to relax the substrate specificity of many DNA polymerases, possibly by allowing enhanced binding of the β- and γ-phosphates of the dNTPs and/or by strengthening weak contacts between polymerases and their DNA substrates. Substitution of adenine with diaminopurine in both template and monomer triphosphates further increases DNA synthesis on GNA templates, emphasizing the requirement for a strong base-pairing interaction between the incoming nucleoside triphosphate and the template for efficient synthesis by polymerases.

A series of structures of a related Bst DNA polymerase (BF) and its complexes with substrates have been solved and provide insight into the possible interactions between Bst DNA polymerase and a GNA template. BF is produced by a different strain of Bacillus Stearothermophilus and shares 86.1% sequence identity (with conserved catalytic and substrate-binding site) with Bst polymerases used. The catalytically active crystal of BF polymerase has provided a detailed molecular picture of primer/template-binding and conformational change during the course of dNTP incorporation. The structure of BF in a complex with a primer/template shows that the single-stranded template forms a nearly 90° turn relative to the duplex region and binds to the O and O1 helices in the finger region. Therefore, the template is not preorganized for simultaneous base-pairing and primer-stacking interactions with the incoming nucleotide. A conserved tyrosine (Y714 of BF) of the O helix orchestrates the conformational change of the nucleotide on the template (N) from a preinsertion site [not stacked against the (N − 1) base pair, the open state] to an insertion site [stacked against the (N − 1) base, the closed state], which is capable of base-pairing with the incoming dNTP. The transition between the open and the closed state for each catalytic cycle involves a large conformational change of the O helix and the translocation of the template by 1 nt. Results suggest that the single-stranded GNA can bind to Bst polymerase in a similar fashion as DNA. The conformational change between the open and closed state only requires the formation of 5 bp upstream from the 3′ terminus of the primer. Therefore, stable duplex formation over a longer region does not appear to be necessary for catalysis. In addition, BF polymerase actively binds to the duplex region of the primer/template by partially unwinding the duplex (up to 4 bp) and forming H-bonding interactions with N3 of purines and O2 of pyrimidines in the minor groove.

Therefore, a B-form-like conformation in the primer/template is not required for the binding of the template strand to polymerases. Results suggest that Bst polymerase could stabilize a short region of heteroduplex formation between the newly synthesized DNA product and the GNA template in the active site to maintain a catalytically active, "closed" conformation. This transient and local enzyme-stabilized base-pairing region would then translocate along the template as DNA synthesis continues.

The structural simplicity and the facile synthesis of GNA make it an attractive system for studying the evolution of functional biopolymers by *in vitro* selection. As a step toward exploring potential biological functions of GNA, at present it has been demonstrated that sequence information can be faithfully copied from GNA to DNA by using DNA polymerases. In addition, the structural resemblance of GNA to DNA/RNA suggests that GNA could be an ancestor or progenitor of the contemporary nucleic acids. GNA monomers can be assembled by condensing a nucleobase with a C3 unit (e.g., glycerol, glyceraldehyde, or glycidol). Previous studies have shown that glycidol can be produced from glycerol carbonate in a reaction catalyzed by zeolites, which suggests that GNA monomers might be formed under prebiotic conditions. Because highly activated GNA monomers such as phosphorimidazolides cyclize rapidly, template-directed primer extension by cyclic monomers or triphosphates deserves further study. Because duplex formation of GNA with itself and with RNA is stereospecific, GNA is also an interesting system in which to study the rise of homochirality during the origins of life. Future studies on the nonenzymatic polymerization of GNA will help shed light on its role as a possible genetic information carrier before the proposed "RNA World."

To summarize; a synthetic molecule called peptide nucleic acid (PNA) combines the information-storage properties of DNA with the chemical stability of a proteinlike backbone.

Drugs based on PNA would achieve therapeutic effects by binding to specific base sequences of DNA or RNA, repressing or promoting the corresponding gene.

Some researchers working to construct artificial life-forms out of mixtures of chemicals are also considering PNA as a useful ingredient for their designs.

PNA-like molecules may have served as primordial genetic material at the origin of life.

.

9 ELECTRICITY, FREQUENCIES, LIFE, AND OUR ENVIRONMENT

Experiments with beams of negative particles were performed in Britain by Joseph John ("J.J.") Thomson (pictured left), and led to his conclusion in 1897 that they consisted of lightweight particles with a negative electric charge, now known as electrons. Thomson was awarded the 1906 Nobel Prize.

The Electron

The word "elektron" in Greek means amber, the yellow fossilized resin of evergreen trees, a "natural plastic material" already known to the ancient Greeks. It was known that when amber was rubbed with dry cloth--producing what one would today call static electricity--it could attract light objects, such as bits of paper.

William Gilbert, a physician who lived in London at the time of Queen Elizabeth I and Shakespeare, studied magnetic phenomena and demonstrated that the Earth itself was a huge magnet, by means of his "terrella" experiment. But he also studied the attraction produced when materials such as amber were rubbed, and named it the **"electric"** attraction. From that came the word **"electricity"** and all others derived from it.

By 1879, a man by the name of Thomas Alva Edison (1847-1931) had produced a new concept: a high resistance lamp in a very high vacuum, which would burn for hundreds of hours. While earlier inventors had produced electric lighting in laboratory conditions, dating back to a demonstration of a glowing wire by Alessandro Volta in 1800, Edison concentrated on commercial application, and was able to sell the concept to homes and businesses by mass-producing relatively long-lasting light bulbs and creating a complete system for the generation and distribution of electricity.

The "**Edison effect**" was the name given to a phenomenon that Edison observed in 1875 and refined later, in 1883, while he was trying to improve his new incandescent lamp. The effect was that, in a vacuum, electrons flow from a heated element -- like an incandescent lamp filament -- to a cooler metal plate. Edison saw no special value in the effect, but he patented it anyway. Edison patented everything in sight. Today we call the effect by the more descriptive term, "*thermionic emission.*"

Now the Edison effect has an interesting feature. The electrons can flow only one way -- from the hot element to the cool plate, but never the other way -- just like the water flow through a check valve. Today we call devices that let electricity flow only one way, diodes.

In 1904, the Edison effect was finally put to use, but not in a light bulb. Radio was in its infancy, and the British physicist John Fleming was working for the British "Wireless Telegraphy" Company. He faced the problem of converting a weak alternating current into a direct current that could actuate a meter or a telephone receiver. Fortunately, Fleming had previously consulted for the Edison & Swan Electric Light Company of London. The connection suddenly clicked in his mind, and he later wrote,

"To my delight I ... found that we had, in this peculiar kind of electric lamp, a solution ..."

Fleming realized that an Edison-effect lamp would convert alternating current to a direct current because it let the electricity flow only one way. Fleming, in other words, invented the first vacuum tube. Of course, most vacuum tubes have been replaced with solid-state transistors today; but they haven't vanished entirely. They still survive, in modified forms, in things like television picture tubes and X-ray sources.

During the 1800s it also became evident that the electric charge had a natural unit, which could not be subdivided any further, and in 1891 Johnstone Stoney proposed to name it "**electron**." When J.J. Thomson discovered the light particle which carried that charge, the name "electron" was applied to it. The many applications of electrons moving in a near-vacuum or inside semiconductors were later dubbed "**electronics**."

The electron itself is a subatomic particle with a negative elementary electric charge. It has no known components or substructure; in other words, it is generally thought to be an elementary particle. An electron has a mass that is approximately 1/1836 that of the proton. The intrinsic angular momentum (spin) of the electron is a half-integer value in units of $\hbar$, which means that it is a fermion. The antiparticle of the electron is called the positron; it is identical to the electron except that it carries electrical and other charges of the opposite sign.

When an electron collides with a positron, both particles may be totally annihilated, producing gamma ray photons. Electrons, which belong to the first generation of the lepton particle family, participate in gravitational, electromagnetic and weak interactions.

Electrons, like all matter, have quantum mechanical properties of both particles and waves, so they can collide with other particles and can be diffracted like light. However, this duality is best demonstrated in experiments with electrons, due to their tiny mass. Since an electron is a fermion, no two electrons can occupy the same quantum state, in accordance with the Pauli exclusion principle.

The Earth's Atmosphere

The atmosphere surrounds Earth and protects us by blocking out dangerous rays from the sun. The atmosphere is a mixture of gases that becomes thinner until it gradually reaches space. It is composed of Nitrogen (78%), Oxygen (21%), and other gases (1%).

Oxygen is essential to life because it allows us to breathe. Some of the oxygen has changed over time to ozone. The ozone layer filters out the sun's harmful rays. Recently, there have been many studies on how humans have caused a hole in the ozone layer.

We are also being sold a story about how humans are affecting Earth's atmosphere through the greenhouse effect. Due to increases in gases, like carbon dioxide, that trap heat being radiated from the Earth, some scientists believe that the atmosphere is having trouble staying in balance creating the greenhouse effect. I personally believe this story to be a fabrication in order to initiate and move forward with geoengineering and aerosol projects that are at the very least blocking out our necessary sunlight, and the worst and most frightening concept is that the aerosol projects are to feed self-assembling and replicating nanotechnology and other synthetic organisms, including assimilation with and augmentation of human beings and other living things as well on a scale that is undetectable to the human eye.

The atmosphere is divided into five layers depending on how temperature changes with height. Most of the weather occurs in the first layer.

Layers of the Earth's Atmosphere

The atmosphere is divided into five layers. It is thickest near the surface and thins out with height until it eventually merges with space.

The **troposphere** is the first layer above the surface and contains half of the Earth's atmosphere. Weather occurs in this layer.

Many jet aircrafts fly in the **stratosphere** because it is very stable. Also, the ozone layer absorbs harmful rays from the Sun.

Meteors or rock fragments burn up in the **mesosphere**.

The **thermosphere** is a layer with auroras and contains the ionosphere. It is also where the space shuttle orbits.

The atmosphere merges into space in the extremely thin **exosphere**. This is the upper limit of our atmosphere.

The Troposphere

The troposphere is the lowest layer of the Earth's atmosphere. The air is very well mixed and the temperature decreases with altitude.
Air in the troposphere is heated from the ground up. The surface of the Earth absorbs energy and heats up faster than the air does. The heat is spread through the troposphere because the air is slightly unstable.

Weather occurs in the Earth's troposphere.

The Stratosphere

In the Earth's stratosphere, the temperature increases with altitude. Theoretically, the release of ozone from Earth causes the increasing temperature in the stratosphere. Ozone is concentrated around an altitude of 25 kilometers. The ozone molecules absorb dangerous kinds of sunlight, which heats the air around them.

The stratosphere is located above the top of the troposphere.

Ozone - An Overview

The Ozone Hole. Pollution. Skin Cancer. Why does the topic of ozone make the news so much? How important is the ozone in our atmosphere? Why are scientists so concerned about its increase near the surface of the Earth and its disappearance higher up in the atmosphere?

First things first - what is ozone? Ozone is made of three oxygen atoms (O3). The oxygen in our atmosphere that we breathe is made up of two oxygen atoms (O2). When enough ozone molecules are present, it forms a pale blue gas. Ozone has the same chemical structure whether it is found in the stratosphere or the troposphere. Where we find ozone in the atmosphere determines whether we consider it to be "good" or "bad"!

In the troposphere, the ground-level or "bad" ozone is an air pollutant that damages human health, vegetation, and many common materials. It is a key ingredient of urban smog.

In the stratosphere however, we find the "good" ozone that protects life on Earth from the harmful effects of the Sun's ultraviolet rays. We have good reason to be concerned about the thinning of the ozone layer in the stratosphere. We also have good reason to be concerned about the buildup of ozone in the troposphere. Although simplistic, the saying "*Good up high and bad nearby*", sums up ozone in the atmosphere.

Ozone in the Stratosphere

About 90% of the ozone in the Earth's atmosphere is found in the region called the stratosphere. This is the atmospheric layer between 16 and 48 kilometers (10 and 30 miles) above the Earth's surface. Ozone forms a kind of layer in the stratosphere, where it is more concentrated than anywhere else.

Ozone and oxygen molecules in the stratosphere absorb ultraviolet light from the Sun, providing a shield that prevents this radiation from passing to the Earth's surface. While both oxygen and ozone together absorb 95 to 99.9% of the Sun's ultraviolet radiation, only ozone effectively absorbs the most energetic ultraviolet light, known as UV-C and UV-B. This ultraviolet light can cause biological damage like skin cancer, tissue damage to eyes and plant tissue damage. The protective role of the ozone layer in the upper atmosphere is so vital that scientists believe life on land probably would not have survived - and could not exist today - without it.

The ozone layer would be quite good at its job of protecting Earth from too much ultraviolet radiation - that is, it would if humans did not contribute to the process. It's now known that ozone is destroyed in the stratosphere and that some human-released chemicals such as CFC's are speeding up the breakdown of ozone, so that there are "holes" now in our protective shield.

While the stratospheric ozone issue is a serious one, in many ways it can be thought of as an environmental success story. Scientists detected the developing problem, and collected the evidence that convinced governments around the world to take action. Although the elimination of ozone-depleting chemicals from the atmosphere will take decades yet, we have made a strong and positive beginning. For the first time in our species' history, we have tackled an environmental issue on a global scale.

The Mesosphere

In the Earth's mesosphere, the air is relatively mixed together and the temperature decreases with altitude. The atmosphere reaches its coldest temperature of around -90°C in the mesosphere. This is also the layer in which a lot of meteors burn up while entering the Earth's atmosphere.

The mesosphere is on top of the stratosphere The upper parts of the atmosphere, such as the mesosphere, can sometimes be seen by looking at the very edge of a planet.

The Thermosphere

The thermosphere is the fourth layer of the Earth's atmosphere and is located above the mesosphere. The air is really thin in the thermosphere. A small change in energy can cause a large change in temperature. That's why the temperature is very sensitive to solar activity. When the sun is active, the thermosphere can heat up to 1,500° C or higher!

The Earth's thermosphere also includes the region of the atmosphere called the ionosphere. The ionosphere is a region of the atmosphere that is filled with charged particles. The high temperatures in the thermosphere can cause molecules to ionize. This is why an ionosphere and thermosphere can overlap.

The Ionosphere

Scientists call the ionosphere an extension of the thermosphere. So technically, the ionosphere is not another atmospheric layer. The ionosphere represents less than 0.1% of the total mass of the Earth's atmosphere. Even though it is such a small part, it is extremely important!

The upper atmosphere is ionized by solar radiation. That means the Sun's energy is so strong at this level, that it breaks apart molecules. So there ends up being electrons floating around and molecules which have lost or gained electrons. When the Sun is active, more and more ionization happens!

Temperatures in the ionosphere just keep getting hotter as you go up!

Different regions of the ionosphere make long distance radio communication possible by reflecting the radio waves back to Earth. It is also home to auroras.

When we consider that the ionosphere surrounding our planet is electrically positive charged, while the earth's surface carries a negative charge, we must conclude that this amounts to a prevailing electrical tension within the earth/ionosphere cavity. This tension is discharged when thunderstorms develop in this cavity. In physics two concentric electrically charged balls, one placed inside the other, are called ball condensers, or capacitors.

The physicist and inventor Nikola Tesla was the first to carry out wireless energy experiments at Colorado Springs, USA, which produced such powerful electrical tensions that they resulted in the creation of artificial lightning. These lightning flashes also produced radio waves. Due to the extremely low frequency these waves could penetrate the earth without resistance and thereby Tesla discovered the resonance frequency of the earth. Unfortunately Tesla was before his time and his discoveries were not taken seriously.

Regions of the Ionosphere

The ionosphere is broken down into the D, E and F regions. The breakdown is based on what wavelength of solar radiation is absorbed in that region most frequently.

The D region is the lowest in altitude, though it absorbs the most energetic radiation, hard x-rays. The D region doesn't have a definite starting and stopping point, but includes the ionization that occurs below about 90km.

The E region peaks at about 105km. It absorbs soft x-rays.

The F region starts around 105km and has a maximum around 600km. It is the highest of all of the regions. Extreme ultra-violet radiation (EUV) is absorbed there.

On a more practical note, the D and E regions reflect AM radio waves back to Earth. Radio waves with shorter lengths are reflected by the F region. Visible light, television and FM wavelengths are all too short to be reflected by the ionosphere. So TV stations are made possible by satellite transmissions.

The Sun's Effect on the Ionosphere

Invisible layers of ions and electrons are found in the Earth's atmosphere. We call this region of atmosphere the ionosphere. The main source of these layers is the Sun's ultraviolet light which ionizes atoms and molecules in the Earth's upper atmosphere. During this process, electrons are knocked free from molecules or particles in the atmosphere.

Flares and other big events on the Sun produce increased ultraviolet, x-ray and gamma-ray photons that arrive at the Earth just 8 minutes later (other particles from the Sun may arrive days later) and dramatically increase the ionization that happens in the atmosphere. So, the more active the Sun, the thicker the ionosphere!

The Exosphere

Very high up, the Earth's atmosphere becomes very thin. The region where atoms and molecules escape into space is referred to as the exosphere. The exosphere is on top of the thermosphere.

Active Signals in Our Environment

More than half a century after Nicola Tesla's artificial lightening experiments; in 1952 the German physicist Professor W.O. Schumann of the Technical University of Munich predicted that there are electromagnetic standing waves in the atmosphere, within the cavity formed by the surface of the earth and the ionosphere. This came about by Schumann teaching his students about the physics of electricity. During a lesson about ball condensers he asked them to calculate the frequency between the inner and outer ball, meaning the earth and ionosphere layer. They came up with a calculation of 10Hz.

This was confirmed in 1954 when measurements by Schumann and König detected resonances at a main frequency of 7.83 Hz. In the years following this discovery, several investigators worldwide have researched "Schumann resonance" and a number of properties and characteristics have now been established.

Schumann Resonance Properties

The spherical earth-ionosphere cavity is created by the conductive surface of the earth and the outer boundary of the ionosphere, separated by non-conducting air. Electromagnetic impulses are generated by electrical discharges such as lightning, the main excitation source, and spread laterally into the cavity. Lightning discharges have a "high-frequency component", involving frequencies between 1 kHz and 30 kHz, followed by a "low-frequency component" consisting of waves and frequencies below 2 kHz and gradually increasing amplitude. This produces electromagnetic waves in the very low frequency (VLF) and extremely low frequency (ELF) ranges.

ELF waves at 3 Hz to 300 Hz are propagated as more or less strongly attenuated waves in the space between the earth and the ionosphere, which provides a waveguide for the signals. Certain wavelengths circumnavigate the earth with little attenuation due to the fact that standing waves are formed within the cavity, the circumference of which is "approximately equal to the wavelength which an electromagnetic wave with a frequency of about 7.8 Hz would have in free space" (König, 1979, p34). It is the waves of this frequency and its harmonics at 14, 20, 26, 33, 39 and 45 Hz that form Schumann Resonances.

On a global scale the total resonant spectrum is the effect of the global lightning worldwide which is estimated at an average of 100 strokes per second. Since there is a concentration of lightning activity during the afternoon in Southeast Asia, Africa and America there are Schumann Resonance amplitude peaks at 10, 16 and 22 UT (universal time), with activity over America around 22 UT being dominant.

There are also +/-0.5 Hz variations in the center frequency, caused by a diurnal increase in ionization of the ionosphere as a result of radiation from the sun, having the effect of reducing the height of the ionosphere at around 12 GMT time. Another factor which influences center frequency is sunspot activity.

Although the existence of the Schumann Resonance is an established scientific fact, there are very few scientists who are aware of the importance of this frequency as a tuning fork for Life. It has been proposed that it is not merely a phenomenon caused by lightning in the atmosphere, but a very important electromagnetic standing wave, acting as background frequency and influencing biological oscillators within the mammalian brain.

At the time when Schumann published his research results in the journal `Technische Physik', Dr Ankermueller, a physician, immediately made the connection between the Schumann resonance and the alpha rhythm of brainwaves. He found the thought of the earth having the same natural resonance as the brain very exciting and contacted Professor Schumann, who in turn asked a doctorate candidate to look into this phenomenon. This candidate was Herbert König who became Schumann's successor at Munich University.

König demonstrated a correlation between Schumann Resonances and brain rhythms. He compared human EEG recordings with natural electromagnetic fields of the environment (1979) and found that the main frequency produced by Schumann oscillations is very close to the frequency of alpha rhythms.

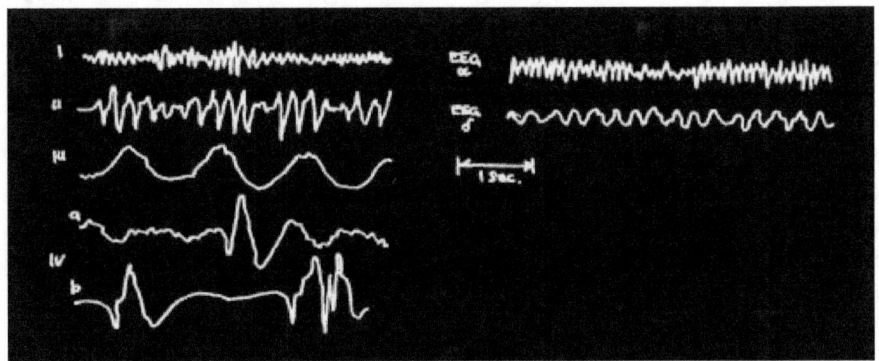

Natural electromagnetic processes in the environment (I-IV), human EEG readings in comparison. Schumann oscillations (I) and the EEG α-rhythm, as well as locally conditioned fluctuations of the electric field (II) and the EEG δ-rhythm, show a noticeable similarity in their temporal variation. From König, 1979.

Dr König carried out further measurements of Schumann resonance and eventually arrived at a frequency of exactly 7.83 Hz, which is even more interesting, as this frequency is one which applies to mammals. For instance, septal driving of the hippocampal rhythm in rats has been found to have a minimum threshold at 7.7 Hz (Gray, 1982).

This relationship has been explored by a number of investigators. One of the foremost researchers in this field is Dr Wolfgang Ludwig, who has been investigating Schumann Resonance and its place in nature for many years.

The Research of Dr Wolfgang Ludwig

It was Dr Wolfgang Ludwig who carried out further measurements while writing his thesis on the Schumann Resonance. His aim was to measure what kind of natural signals actually exist in a healthy environment.

He became aware of the fact that due to manmade electromagnetic signals within the atmosphere, the accurate measurement of Schumann waves was almost impossible in the city. For this reason he decided to take measurements out at sea where, due to good electrical conductivity, the Schumann waves are stronger. He then had the idea to take underground measurements in mines. Here he recognized that the magnetic field of the earth fluctuated too. This was also investigated by Dr Robert Becker in his book `Electricity and Vitality: The spark of Life'.

Dr Ludwig came up with an excellent idea to take accurate measurements. When taking measurements at the earth's surface, the reading is the result of two signals, one coming from above and one from below. But subsequently taking measurements below ground makes it possible to come up with exact readings by separating the two.

Dr. Ludwig discovered that an imbalance of Schumann and Geomagnetic waves induces micro-stress. The first Schumann waves generators were developed by Dr. Ludwig and colleagues in 1974 and have been in a continuous state of development ever since. On the basis of the research work of Dr. Ludwig the NASA constructed "Schumann and Geomagnetic" frequency generators in its manned satellites. The classical Chinese medicine has traditionally said that man needs environmental signals of two kinds - yang input from above (Schumann waves) and yin input from below (geomagnetic waves). Both of these, it is said, must be in balance.

Yin and Yang

During his research Dr Ludwig came across the ancient Chinese teachings which state that Man needs two environmental signals: the YANG (masculine) signal from above and the YIN (feminine) signal from below. This description fits the relatively strong signal of the Schumann wave surrounding our planet being YANG and the weaker geomagnetic waves coming from below, from within the planet, being the YIN signal

The Chinese teachings state that to achieve perfect health, both signals must be in balance. Dr Ludwig found that this is indeed the case. He writes in his book `Informative Medizin' that research carried out by E.Jacobi at the University of Duesseldorf showed that the one sided use of Schumann (YANG) wave simulation without the geomagnetic (YIN) signal caused serious health problems. On the other hand, the absence of Schumann waves creates a similar situation. Professor R.Wever from the Max Planck Institute for Behavioural Physiology in Erling-Andechs, built an underground bunker which completely screened out magnetic fields. Student volunteers lived there for four weeks in this hermetically sealed environment. Professor Wever noted that the student's circadian rhythms diverged and that they suffered emotional distress and migraine headaches. As they were young and healthy, no serious health conditions arose, which would not have been the case with older people or people with a compromised immune system. After only a brief exposure to 7.8 Hz (the very frequency which had been screened out), the volunteers health stabilized again.

The same complaints were reported by the first astronauts and cosmonauts, who, out in space, also were no longer exposed to the Schumann waves. Now modern spacecrafts are said to contain a device which simulates the Schumann waves.

All the aforesaid points to the fact that the ancient teachings are correct. Mankind depends on two subtle environmental signals, the Yin from below and the Yang from above.

There is an urgent need for further public research into the Schumann Resonance Effect.

Although Schumann Resonance could easily be confirmed by measurements at the time of its discovery, it is no longer so obvious due to our atmosphere being filled with manmade radiation noise at different frequencies. This is almost drowning out the natural signals - signals that have been there through aeons of evolution. It is possible that these signals act like a natural tuning fork, not just for the biological oscillators of the brain, but for all processes of life.

With the advent of new wireless technology, in particular microwaves pulsed at frequencies close to Schumann Resonance as in mobile telephony, another threat is emerging. We may be creating an environment that is literally 'out of tune' with Nature itself. And it is at this point that there is an urgent need for us to understand how everything alive responds to the most subtle changes in magnetic and electromagnetic fields surrounding us. For instance, we need to examine the possible interaction between magnetite crystals within cells and manmade magnetic fields in the environment.

There is a great need for independent research into the bio-compatibility between natural and manmade signals. By linking together the potential importance of Schumann Resonance and the dangers posed by manmade pulsed frequencies, it will become apparent that unless we find a way to use bio-compatible signals to power new technology, we may expose all life to dangers previously not encountered. We may have to pay a high price for this shortsightedness. Serious attention must now be paid to the possible biological role of standing waves in the atmosphere, so that we do not overlook the importance of oscillations in nature that may be central to consciousness and life itself.

The late Dr Neil Cherry, a fierce opponent of the frequencies used in mobile telephony, has also focused on the importance of Schumann Resonance in his publications *'Schumann Resonances, a plausible biophysical mechanism for the human health effects of Solar/Geomagnetic Activity'* (2002), and *'Human intelligence: The brain, an electromagnetic system synchronized by the Schumann Resonance signal'* (2003).

If organisms do in fact respond to, and perhaps depend on, electromagnetic fields as weak as that produced by Schumann resonance at 0.22-1.12 mV/m (from Cherry, 2002), this is of major significance for the development of present and future wireless technologies.

Three kinds of bioenergetically active signals are present in the earth´s natural environment: Schumann waves, geomagnetic waves and solar waves.

Schumann waves

At about 100 kilometers above the surface, the earth is surrounded by the ionosphere. The vast space between the ionosphere and the earth´s surface acts as an enormous electromagnetic resonance cavity, like the sound box of a musical instrument. These resonating electromagnetic waves are known as Schumann waves, after their discoverer, the physicist W.O. Schumann.

The basic waveform (first harmonic) has a frequency of 7,8 Hz. Exactly the same frequency is present in the main control center of the human brain, the Hippocampus / Hypothalamus; (i.e. the region of our brain which is important for attentiveness and concentration).

Geomagnetic waves

The earth´s crust contains 64 elements which are vital as the so-called "trace elements". Each of these trace elements possesses its own characteristic types of vibrations. The earth´s magnetic field is influenced or modulated by these vibrations, the resulting "modulation" being known as geomagnetic waves.

The earth crust contains the same essentially vital mineral material ("trace elements") as those existing in the red blood corpuscles of humans. The relation among one another is nearly the same.

Solar waves

Light affects our hormone household, the metabolism and has effects on our immune system.

While in former times humans spent the largest part of their daily life in the free, most humans today spend their time in dwellings, factories, offices or schools.

The absence or blocking of these natural and vital environmental signals can affect the general condition and regulation of our bodies and of all life across the entire planet.

Extremely Low Frequency Magnetic Fields

Clinical research indicates that somewhere between 25% to 75% of human and animal subjects may exhibit marked psychophysiological sensitivity to extreme low frequency magnetic and electric fields.

Brainwave entrainment can be demonstrated electroencephalographically when subjects are in the vicinity of oscillations in the frequency range of approximately 3 to 20 Hz at intensities of below 100 nanoteslas. (1 tesla = 10^4 gauss)

ELF fields of 6.67 Hz, 6.26 Hz and lower tend to produce symptoms of confusion, anxiety, depression, tension, fear, mild nausea and headaches, cholinergia, arthritic-like aches, insomnia, extended reaction times, hemispheric EEG desynchronization, and many other vegetative disturbances.

H and B field (magnetic vector) oscillations of 7.8, 8.O and 9 Hz produce anxiety-relieving and stress-reducing effects mimicking some "meditative" states. It has been speculated that frequencies in this range may be the universally permeating "clock frequencies" or carrier signals on which "mind" or "consciousness" states can be impressed and interact with other life-forms in the nebulous realms of ESP, psychotronics, distant healing, radiesthesia, and related paranormal but anecdotal phenomena.

Coherent ELF energies have the unique and interesting property of almost lossless propagation within the earth-ionosphere cavity waveguide, and attenuation of these signals due to distance from transmitter sites is negligible. Power losses are O.8 db per Mm (million meters) . The magnetic vectors, unlike electrical (E-wave) components, permeate any substance and cannot be effectively shielded, even by iron, mu-metal, lead, copper, "Faraday cages", etc.

The established physics of radio propagation therefore suggests that vast geographical areas can be readily mood-manipulated by transmissions of EM energy within the earth-ionosphere cavity wave guide. The accidental discovery in 1938 of the "Luxembourg Effect", by which a transmitter of relatively low power, fed to an antenna array producing a circularly-polarized signal, can "piggy-back" to any preexisting RF energy source and produce disturbances of the ionosphere at any frequency desired (which then "modulates" all other energies within that "waveguide" is fully described in the literature of 40 years ago but probably forgotten or overlooked by today's technologists.

Piezoelectric Effect

Crystals which acquire a charge when compressed, twisted or distorted are said to be piezoelectric. This provides a convenient transducer effect between electrical and mechanical oscillations. Quartz demonstrates this property and is extremely stable. Quartz crystals are used for watch crystals and for precise frequency reference crystals for radio transmitters. Rochelle salt produces a comparatively large voltage upon compression and was used in early crystal microphones. Barium titanate, lead zirconate, and lead titanate are ceramic materials which exhibit piezoelectricity and are used in ultrasonic transducers as well as microphones. If an electrical oscillation is applied to such ceramic wafers, they will respond with mechanical vibrations which provide the ultrasonic sound source. The standard piezoelectric material for medical imaging processes has been lead zirconate titanate (PZT). Piezoelectric ceramic materials have found use in producing motions on the order of nanometers in the control of scanning tunneling microscopes.

The word piezo is Greek for "push". The effect known as piezoelectricity was discovered by brothers Pierre and Jacques Curie when they were 21 and 24 years old in 1880.

There is a magnetic analog where ferromagnetic material respond mechanically to magnetic fields. This effect, called magnetostriction, is responsible for the familiar hum of transformers and other AC devices containing iron cores.

The Next Battlefield; Your Mind

We are constantly being assaulted; from having our brainwaves deliberately changed en masse by transmitters regulating our state of consciousness, to how we are victims of electromagnetic waves disrupting the state of our health and finally how many of us will die, as decided by our global masters.

Earth is wrapped in a donut shaped magnetic field. Circular lines of flux continuously descend into the North Pole and emerge from the South Pole. The ionosphere, an electromagnetic-wave conductor, 100 km above the earth, consists of a layer of electrically charged particles acting as a shield from solar winds. Natural waves are related to the electrical activity in the atmosphere and are thought to be caused by multiple lightning storms.

Collectively, these waves are called "The Schumann Resonance,' the current strongest at 7.8 Hz.

These are quasi-standing extremely low frequency (ELF) waves that naturally exist in the earth's "electromagnetic" cavity, the space between the ground and the ionosphere. These "earth brainwaves" are identical to the spectrum of our brainwaves. (1 hertz = 1 cycle per second, 1 Khz = 1000, 1 Mhz =1 million. A 1 Hertz wave is 186,000 miles long; 10 Hz is 18,600 miles. Radio waves move at the speed of light.)

Living beings are designed (or have evolved... depending on your beliefs) to resonate to this natural frequency pulsation in order to evolve harmoniously. The ionosphere is being manipulated by US government scientists using multiple Alaskan transmitters called HAARP, (High-Frequency Active Auroral Research Program) which sends focused radiated power to heat up sections of the ionosphere, which bounces power down again.

ELF waves from HAARP, when targeted on certain areas, can engineer weather and create mood changes affecting millions of people.

The intended wattage is 1,700 billion watts of power. A former govt. insider deduced they want to flip the world upside down. Sixty-four (64) elements in the ground modulate, with variation, the geomagnetic waves naturally coming from the ground. The "earth's natural brain rhythm" above is balanced with these.

These are the same minerals as the red blood corpuscles. There is a relation between the blood and geomagnetic waves. An imbalance between Schumann and geomagnetic waves disrupts biorhythms.

These natural geomagnetic waves are being replaced by artificially created very low frequency (VLF) ground waves coming from GWEN Towers.

What Are The GWEN Towers?

GWEN (Ground Wave Emergency Network) transmitters, placed 200 miles apart across the USA, allow specific frequencies to be tailored to the geomagnetic-field strength in each area, allowing the magnetic field to be altered.

They operate in the VLF range, with transmissions between VLF 150 and 175 KHz. They also emit UHF waves of 225 - 400 MHz.

The VLF signals travel by waves that hug the ground rather than radiate into the atmosphere. A GWEN station transmits up to a 300-mile radius, the signal dropping off sharply over distance. The entire GWEN system consists of (depending on source of data), from 58 to an intended 300 transmitters, spread across the USA, each with a tower 299-500 ft high.

Three hundred (300) ft. of copper wire fans out in a spoke like fashion from the base of the underground system, interacting with the earth like a thin shelled conductor, radiating radio wave energy for very long distances through the ground. The USA bathes in this magnetic field which rises to 500 ft, even going down to basements, so everyone is subject to mind control.

The whole artificial ground wave spreads out over USA like a web. It is easier to mind-control and hypnotize people who are bathed in an artificial electromagnetic wave.

GWEN transmitters have many different functions, including controlling the weather, mind, behavior and mood control of the populace.

They are also used to send synthetic telepathy disguised as infrasound to those victims of US government mind-control implants.

These towers work in conjunction with HAARP and the Russian Woodpecker transmitter, a system similar to HAARP. The Russians openly market a small version of their weather-engineering system called Elate, which can fine-tune weather patterns over a 200 mile area and have the same range as the GWEN unit. One such system operates at the Moscow airport.

The GWEN Towers shoot enormous bursts of energy into the atmosphere in conjunction with HAARP. The internet website: www.cuttingedge.org, published an expose on how the major floods of 1993 in the Mid-Western United States were instigated by these systems.

Invisible, enormous rivers of water, consisting of vapors that flow, move towards the poles in the lower atmosphere. They rival the flow of the Amazon River and are 420 to 480 miles wide and up to 4,800 miles long.

They are 1.9 miles above the earth and move 340 lbs of water per second. There are 5 atmospheric rivers in each Hemisphere. A massive flood can be created by damming up one of these massive vapor rivers, causing huge amounts of rainfall to be dumped. The GWEN Towers positioned along the areas north of the Missouri and Mississippi Rivers were turned on for 40 days and 40 nights, probably mocking the Flood of Genesis. (This was in conjunction with HAARP).

The damming of the vapor rivers creates a river of electricity flowing thousands of miles through the sky and down to the polar ice-cap, manipulating the jet-stream. Again, these two major rivers flooded, causing agricultural losses of $12-15 billion. HAARP also produces earthquakes by focusing on the fault lines. GWEN Towers are positioned on the fault lines and volcanic areas of the Pacific Northwest.

In 1963, Dr. Robert Becker explored effects of external magnetic-fields on brainwaves, showing a relationship between psychiatric hospital admissions and solar magnetic storms. He exposed volunteers to pulsed magnetic-fields similar to magnetic storms, and found a similar response. In the United States, sixty (60) Hz electric-power ELF waves vibrate at the same frequency as the human brain. In the United Kingdom, fifty (50) Hz electricity emissions depress the thyroid.

Dr. Andrija Puharich (in the 1950 & 60s), found that a clairvoyant's brainwaves turned to 8 Hz when their psychic powers were operative. In 1956, he observed an Indian Yogi controlling his brainwaves, deliberately shifting his consciousness from one level to another. Puharich trained people via bio-feedback to do this consciously, that is, creating 8 Hz waves with the technique of bio-feedback. A psychic healer generated 8 Hz waves through a hands-on healing process, actually alleviating that patient's heart trouble; the healer's brain emitting 8 Hz.

One person, emitting a certain frequency, can make another also resonate to the same frequency. Our brains are extremely vulnerable to any technology that sends out ELF waves, because they immediately start resonating to the outside signal by a kind of tuning-fork effect.

Puharich further experimented, discovering that, 7.83 Hz (earth's pulse rate) made a person "feel good," producing an altered-state.

10.80 Hz causes riotous behavior.

6.6 Hz causes depression.

Puharich made ELF waves change RNA and DNA in the body, breaking hydrogen bonds to make a person resonate at a higher vibratory rate.

He really wanted to go beyond the psychic 8 Hz brainwave and attract psi phenomena.

James Hurtak, who once worked for Puharich, also wrote in his book 'The Keys of Enoch' that ultra-violet caused hydrogen bonds to break and this raised the vibratory rate.

Puharich presented the mental effects of ELF waves to military leaders, but they would not believe him. He then gave this information to certain dignitaries of other Western nations. The US Government burned down his home in New York to shut him up, whereas he then fled to Mexico. However, the Russians discovered which ELF frequencies affected what portion of the human brain; it was on July 4, 1976, that they began zapping the U.S. Embassy in Moscow with electromagnetic-waves, varying the signal, also focusing on 10 Hz. (10 Hz puts people into a hypnotic state).

Russians and North Koreans use this in portable mind-control machines to extract confessions. (This system can also be found in some American Churches to help the congregation believe!)

This Russian "Woodpecker" signal was traveling across the world from a transmitter near Kiev. The US Air Force identified 5 different frequencies in this compound that the harmonic Woodpecker was sending through the earth and atmosphere.

In 1901, Nikola Tesla, Nobel prize winner in Physics (shared with Einstein) revealed that power could be transmitted through the ground using ELF waves. Nothing stops or weakens these signals. The Russians retrieved Tesla's papers when they were returned to Yugoslavia after his death.

In Mexico, Puharich continued to monitor the Russian ELF wave signal and the higher harmonics (5.340 MHz) in the MHz range. He was somehow induced to work for the CIA and he and Dr. Robert Becker designed equipment to measure these waves and their effect on the human brain. Puharich started his work by putting dogs to sleep.

By 1948/49, he had graduated to monkeys, deliberately destroying their eardrums to enable them to pick up sounds without the eardrum intact.

He discovered a nerve from the tongue could be used to facilitate hearing. He created the tooth implant that mind-control victims are now claiming was put in by their dentist, unbeknownst to them, and causing them to hear "voices in their head.' These were placed under caps or lodged in the jaw.

Implants are now smaller than a hair's width and are injected with vaccine and flu shots. Millions have had this done unknowingly. These "biochips' circulate in the bloodstream and lodge in the brain, enabling the victims to hear "voices' via the implant. There are many kinds of implants now, and it is estimated that 1 in 40 people are recipients of these tiny implants due to alien abductions. However, others have suggested that one in 20 might just be a more accurate statistic.

The fake alien abduction - revealed by many victims - are actually engineered by the U.S. military, using advanced technology to create holograms (4th dimensional pictures) or holographic spaceships outside.

This holographic, advanced technology can actually create a scenario whereas the person believes he/she is going into a spaceship. However, once inside, the aliens are in masquerade; they are actually military personnel outfitted in full costumes, masks et al..

Certainly real abductions occur, however, the "alien abduction" scenario has been most useful to the military in confusing the overall issue. This clouding tends to halt any further investigation into a government participation and inevitably absolves them of any accountability. They are banking on the poor helpless victims feeling too intimidated to reveal such a shocking episode, lest ridicule be visited upon them.

Are the global masters forcing us to respond to an artificially induced vibratory rate?

Those power mongers who want this planet to have a sudden leap in evolution, populated only by the psychically aware and therefore superior class of human? What about the billions of people who are commonly referred to as "useless eaters"; are they to be conveniently disposed of by electro-magnetically-induced cancers and diseases? It certainly causes one to stop and ponder this catastrophic situation.

The physics and engineering behind electromagnetic disease transmission are frightening. Diseases can be reproduced as "disease signatures" in that the vibration of a disease can be manufactured and sent on to be artificially induced. (The brainwave pattern of hallucinogenic drugs can also be copied and sent by ELF waves to induce "visions").

Once diseases are sprayed in the air, electromagnetic waves attune to the disease by using harmonics and sub-harmonics, which in turn make them even more lethal and infectious; actually a more apt description would be deadly, as in inducing death.

The skies are filled daily with chemtrails, those crisscrossed white patterns that are sprayed out across the heavens in the United States and other countries. Are these like contrails that jets emit behind them? Not exactly... contrails dissipate rather quickly, but the chemtrails - those feathery streaks that linger - are deliberately being sprayed and contain insidious chemicals (retrieved, analyzed and proven) which affect the state of consciousness, producing apathy.

This is only one "program" that has been initiated to keep the populace in a continual apathetic state. Add to this, the fluoridation of the drinking water, aspartame, and other highly-questionable drugs.

Fluoride disables the willpower section of the brain, impairing the left occipital lobe. Both fluoride and selenium (in additional amounts) can produce strange effects; one common symptom is that of "hearing voices'.

ELF waves create disturbances in the biological processes of the body, activated on a large scale once the body has been exposed to the aforementioned disease-causing chemtrails.

Some chemtrails have been analyzed under laboratory conditions, the elements shown to cause cleavages in spatial perceptions, blocking the interaction of various amino acids that relate to higher-consciousness. Some were also shown to increase dopamine in the brain thereby producing a listless, euphoric state of lower reactive mind.

This is done to basically create confusion, rendering a person unable to differentiate between the real and illusionary. In addition, some of these chemtrails could be connected with the many UFO abductions occurring on a global massive scale.

Many victims, some recalled under hypnotic regression, have witnessed other abductees laid out on tables (in a sort of assembly line operation) and in the process or being implanted.

Intelligence agencies are in league with each other, behind this disablement of the masses to such a degree where they can't even fight back. In order to implement their plans, that of total control of the populace, they need the overall "frequency" of each victim to function at a specific rate, below the threshold of awareness.

Could this be part of a greater plan with mind-control transmitters covering the whole of USA and England, cleverly disguised as cell phone towers and trees? The power from microwave towers may be turned up to such a level that people will die.

A brain functioning at beta-level (above 13 Hz) is agitated and cannot change the perceptions if it is artificially stabilized to that frequency by technological methods. This frequency may also increase body electricity in others, giving them psychic powers. Is this linked to the new-agers claim of a 12-14 Hz Schumann Resonance, inching us towards the 4th dimension?

Stimulants ingested globally from higher-caffeine, genetically modified plants, may also make an impact on the "global-brain" in the ionosphere that is collecting our brainwaves.

New-age channelers say we are going into a 4th dimensional frequency. They "heard' the voice of some "ET" who informed them of this.

However, some "ETs" are just plain Earthlings in disguise. Using Tesla Technology, Prisoners in the Utah State Prison were bombarded with voices from a "purported" ET, each prisoner receiving the same identical message. Curious, indeed. Today, it is relatively simple to produce these "voices in the head." Implants/microchips are no longer necessary.

In 1988, an inmate in Draper Prison, Utah by the name of David Fratus wrote:

"I began to receive or hear, high-frequency tones in my ears. When I plugged my ears, the tones were still inside and became amplified. It's as if they had became electrified echo chambers with the sounds coming from the inside out.

I then began to hear voices, right in my inner ears and just as vividly as though I were listening to a set of stereo headphones. The end result is that I am now having my brain monitored by an omnipotent computerized mind reading or scanning machine of some sort."

Hundreds of inmates at the Gunnison Facility of the Utah State Prison and the State Hospital were subjected to this brand of mind-control, used as test subjects like rats in a lab. In the early 1970s, this was revealed in the Utah U.S. District Court.

While incarcerated, these inmate test subjects, having been subjected to this Tesla-wave mind-control, tried to seek restitution in the courtroom. Unfortunately, they were unsuccessful.

The University of Utah researched how Tesla-waves could be used to manipulate the mind into hearing voices, overriding and implanting thoughts into the mind, in addition to reading thoughts. They also went about developing eye-implants. Cray (The Cadillac of computers, ultra sophisticated) computers, using artificial intelligence, monitor the victims of government produced implants, sending pre-recorded sound bites or occasional live messages.

They are picked up by satellite and relayed to whatever large TV broadcasting antenna, GWEN tower or other antenna that is nearest the victim. It is believed that some types of implants pick up the signal and broadcast the correct Tesla-wave pattern to create voices within the victim.

The tracking implant keeps the staff and the satellite system informed every few minutes as to exactly where to send the voice signals. The master computer and central HQ for this is reported to be Boulder, CO. It is rumored that transponders are being made there.

The central cellular computer is in the Boulder, CO National Bureau of Standards building. AT&T is also cooperating; several agencies work together on this.

Tim Rifat of UK wrote that "this inter-cerebral hearing" is used to drive the victim mad, as no one else can hear the voices transmitted into the brain of the target. Transmission of auditory data directly into the target's brains using microwave carrier beams is now common practice. Instead of using excitation potentials, one uses a transducer to modify the spoken word into ELF audiograms that are then superimposed on the pulse modulated microwave beam.

On March 21, 1983, The Sydney Morning Herald published the following by Dr. Nassim Abd El-Aziz Neweigy, Assistant Professor of Agriculture, Moshtohor Tukh-Kalubia, Egypt. This article stated:

"Russian satellites, controlled by advanced computers, can send voices in one's own language, interwoven into natural thoughts. They can target the population of choice with this diffused artificial thought process. The chemistry and electricity of the human brain can be manipulated by satellite and even suicide can be induced.

Through ferocious, anti-humanitarian means, the extremist groups are fabricated, the troubles and bloody disturbances are instigated by advanced tele-means via Russian satellites in many countries in Asia, Africa, Europe and Latin America."

Another source says that these have been fed with the world's languages and synthetic telepathy will reach into people's heads making people believe God is speaking to them personally to enact the Second Coming, complete with holograms!

The Russians broke the genetic code of the human brain. They worked out 23 EEG band-wave lengths, 11 of which were totally independent. So if you can manipulate those 11 you can do anything. NSA's (U.S. National Security Agency) Cray computers can remotely track people just by knowing the specific EMF waves (evoked potentials from EEGs in the 30-50 Hz, 5 milliwatt range) of a person's bioelectric-field. Each person's emissions are unique and they can remotely track someone in public.

Now if this isn't a horrifically frightening thought, I don't know what is.

Evoked potentials officially do not exist in physics, but in 1873 a Scotsman, James Clerk Maxwell, discovered that electromagnetic waves have 3 components. He discovered waveforms that exist at a certain number of right-angled rotations away from the electromagnetic field. These are hyper-spacial components, not subject to constraints of time and space.

He claimed that electromagnetic radiation waves were carried by the ether and the ether was disturbed by magnetic lines of force. The hidden component is called only "potential' now and not normally used except for covert hyper-dimensional physics and to manipulate consciousness itself via electromagnetic-waves covering vast areas of the planet.

Approximately one person in 3000 is sensitive to this magnetic waveform component, the telepathic types, (according to a writer called "Majix"), and we are all capable of tuning into this magnetic component by tuning our subconscious to it. Maxwell's successors thought potentials were akin to mysticism because they believed fields contain mass which cannot be created from nothing.

This is what potentials are - both literally and mathematically - an accumulation or reservoir of energy, consequently this hasn't been taught in mainstream physics.

Subliminal words (in the correct electromagnetic-field and attuned to the human brain) that express human consciousness can enter our minds at a subconscious level. Apparently, our brain activity patterns can be measured and stored on super-computers. If a victim needs to have subliminal thoughts implanted, all that is necessary is to capture that brain activity pattern, (saved on the mega computer) and target or match up that person's pattern.

The targeted or specified person is then sent low frequency subliminal messages that they actually think are their own thoughts.

The researcher Majix says our brains are so sensitive, that they are like liquid crystal in response to the magnetic component of the earth. We are sensitive to earth's magnetic changes, changes in the ionospheric cavity and resonate those frequencies ourselves. We are incredibly complex entities, beyond the layperson's comprehension. Our brains are indeed a type of bio-cosmic transducer.

Physicists in Russia have conducted in-depth studies on the effects of the mean annual magnetic-activity, electro-magnetic and electro-static fields on human behavior and the physical body. These electromagnetic and electro-static fields can be likened to what is popularly known as biorhythms. These magnetic frequencies can be manipulated from a very simple piece of equipment operated at extremely low power levels; our brain waves can mimic magnetic frequencies.

From half a second to 4 seconds later, the neurons and brain waves are driven exclusively by this device; power levels almost nonexistent.

All one needs is a circularly polarized antenna, aimed up at the ionospheric cavity and they can then manipulate the moods of everyone within a 75 sq. mile area. The body picks up these "new" manipulated waves and begins to correspond immediately. What is known as the "sleep" frequency will make everyone become tired and sleep.

In Let's Talk MONTAUK, Joyce Murphy presented data that showed that experiments on the 410-420 MHz cycle have been done which could affect the "window frequency to the human consciousness" as a whole.

Preston Nichols, previously mentioned herein, learned from his experimentation with his radio equipment that whenever a 410-420 MHz cycle appeared on the air, a psychic's mind would be "jammed," finally tracing the signal to Montauk Point and the red and white radar antenna on the AF Base there."

In Encounter in the Pleiades by Peter Moon and Preston Nichols, Nichols wrote that,

"Dr. Nicholas Begich, an expert of HAARP, has picked up 435 MHz signals connected to HAARP and that a mind control function is currently

being employed. He claims that 400-450 MHz is the window to human consciousness because it is our present day reality's background frequency."

Tim Rifat wrote in his Microwave Mind Control in the UK article that cellular phones use 435 MHz.

The United Kingdom police use 450 MHz exclusively. Dr. Ross Adey used this frequency for CIA behavioral modification experiments. Police have a vast array of antennae to broadcast this frequency all over UK. Adey used 0.75mW/cm2 intensity of pulse modulate microwave at a frequency of 450 MHz, with an ELF modulation to control all aspects of human behavior. 450 MHZ radar modulated at 60 Hz greatly reduced T-lymphocyte activity to kill cultured cancer cells.

A study in the USA of their 60 Hz power lines repeated this.

Through much study and analysis on this varied topic, independent scientists have concluded that HAARP is slicing up the ionosphere --- the world-brain --- like a microwave knife, producing long tear incisions and destroying the membrane that holds the reservoir of data accumulated of all earth's history.

However, there can be hope if we are aware of all the possibilities that exist. A healer called Mr. A claimed to have received "Ancient Wisdom" from the earth's protective Magnetic-Ring of energy which stores within it all knowledge since time began. Ruth Montgomery wrote about this healer in Born To Heal.

He claimed that if our energy flow is cut off from this magnetic field, (the protective atmospheric magnetic-ring) then the Universal Supply is obstructed and we are no longer in tune with these advantageous frequencies, therefore we begin to get sick.

The Power from this travels in split-seconds around the world and is available to anyone who is capable of receiving and handling it.

The waves from The Ring were automatically translated into words in the healer's mind and interpreted as wisdom to diagnose and heal others; this ability coming from the storehouse of knowledge that has been present since the beginning of time. By tapping into this storehouse, he produced instant miracles, knitting broken bones and removing arthritis. A photo was produced that displayed forked lightning emitting from his fingers.

Trent Goodbaudy

10 CONNECTING THE DOTS

So, just how big is the big picture? Usually people think that if they gain a new perspective from outside the box they will gain a better perspective of what they are trying to understand. Many also believe that this settles at the level of the elite. They want all the power, they want all the control. We need to look at the bigger picture, and I would like you to follow along with me as I connect the dots. When you begin to connect the dots, you will begin to notice a shape take form. The terrain is huge, this shape is what we need to be aware of. The shape that will manifest will be much, much bigger than we have ever dreamed of.

This really is about impressions and perspective.

"The real voyage of discovery consists not in seeking new landscapes ... but in having new eyes."

~Marcel Proust, 1871-1922

Here is a photograph:

And here is an impression:

I am going to give you the impressions that have been made on me by the information I have gathered, and see what you think of what I have found.

Were getting new impressions of the sky. Once upon a time, clouds used to be puffy.

Today, you see a lot of streaks, feathers, along with unnatural criss-cross configurations. When you look closely at what forms your impression, you will see lines and ropes in the sky made by planes. These lines in the sky are forming a "web" around the planet; a haze. Even NASA now admits to "man-made" clouds; they call them "jet-cirrus" clouds. We are told that "new" atmospheric conditions are causing jet contrails to linger in the sky. Anything that stays in the sky for hours fans out, and forms a layer of material thick enough to block sunlight for miles, is not condensation. The "con" in contrail is for condensation, so if it isn't condensation; meaning water droplets that under normal circumstances quickly evaporate... then what is it?

Contrail = "condensation" trail (water vapor)
Why doesn't this water vapor go away ... evaporate?

Ropes in the sky, made by planes forming a haze.

Contrails typically are formed by jets, so that is high-altitude, and low-humidity. Clouds typically form at a low-altitude, in high-humidity. So the rule is this; clouds and contrails need opposite conditions to form.

Contrails   **High-Altitude**
                Low-Humidity

Clouds   **Low-Altitude**
                High-Humidity

Today, we have vapor trails or "condensation-trails" turning into clouds and spreading out to form a low altitude white haze.

In many parts of the world we are losing our blue skies, you see --- in the past, and --- in the present. Many people even have a phrase for this condition, called "white-out".

If you travel in a plane, this is what you will see from the window. There is a demarcation line, Sophia Smallstrom calls it the "sprayline", below which the world is seen through a veil, and above which you have this deep blue sky. We live in the veil. We live in a white world of whatever it is that is descending on us from above. Our visibility is impaired, the sky as we look up at it is a much lighter blue. The change has been so incremental that most of us are too busy to have noticed.

Airborne Environmental Samples and Soil/Water samples reveal 3 types of material:

- **Metallic salts (oxides)**
- **Filaments / Fibers**
- **Engineered Biologicals**

When you look at airborne environmental samples and the soil and water samples that people have collected throughout the country, you get three types of materials; metallic salts (otherwise known as oxides), filaments or fibers, and engineered biologicals. Metallic salts are oxides; aluminum, titanium, and barium, are three examples of this type of material. It appears now that the air around us has been filled with these metallic oxides, which are conductive. Conductive means that... the air is no longer neutral. Air is supposed to be neutral to support life, it is not supposed to be charged.

The air around us is also filled with fibers and if you get a black light (available at your local hardware store), you can see a particle storm of airborne fibers floating around and flying in the air.

How do you make clouds and rain?

Condensation Nuclei

Dust or salt particles + Evaporated Moisture = Clouds

Clouds are formed from moisture that evaporates from the earth's surface, carried upward on air that is warmed by the sun. As the warm air rises, it begins to cool. There is a certain point, called the "dew-point", at which point the moisture in the air begins to condense around particulates which are in the air. These particulates are usually dust or salt, and are called "*condensation nuclei*". They have an affinity for moisture. As more water vapor gathers around the particulates, the combination becomes unstable and produces a raindrop.

So, we know that particulates in the atmosphere nucleate (gather) into raindrops.

Daniel Rosenfeld, PhD - An atmospheric Scientist at Hebrew University, Jerusalem wrote an article with the headline - "*Pollution Stops Rain*" - In this report, Dr Rosenfeld states; "*Pollution particulates actually prevent rain, by forming very tiny, very stable droplets that remain suspended and do not fall earthward.*"

If this is correct, then it suggests that the smaller the particulate, the more stable the droplet. This is how you would get clouds to hang in the sky as haze (or so called "clouds" anyway).

It would seem that you could create cloud-like masses, simply by releasing small enough particulates into the atmosphere. These clouds would not fall, they would spread outward instead to form haze. This leads us to the subject of geoengineering.

Geoengineering

Geoengineering is the large scale manipulation of the earth's environment to suit human needs and promote habitability. This was initially intended as a solution for man-made global warming and climate change.

A few years ago scientists began to discuss something called SAG Stratospheric Aerosol Geoengineering which was supposedly spraying soot and sulfates in the atmosphere to form a shield against the heat of the sun.

This presumes the notion of man-made global warming, and the idea that man-made science can "rescue us" from climate change, which is natures retaliation against us for creating global warming.

Now we being told that this is a future consideration, but for years we have already been seeing strange clouds and hazy skies. What exactly is falling on us and being sprayed on us?

They are telling us that there are new forms of clouds like this hooked cloud, called the "Cirrus Uncinus". Or the sky filling "asperatus" (below), which causes a blanket canopy effect.

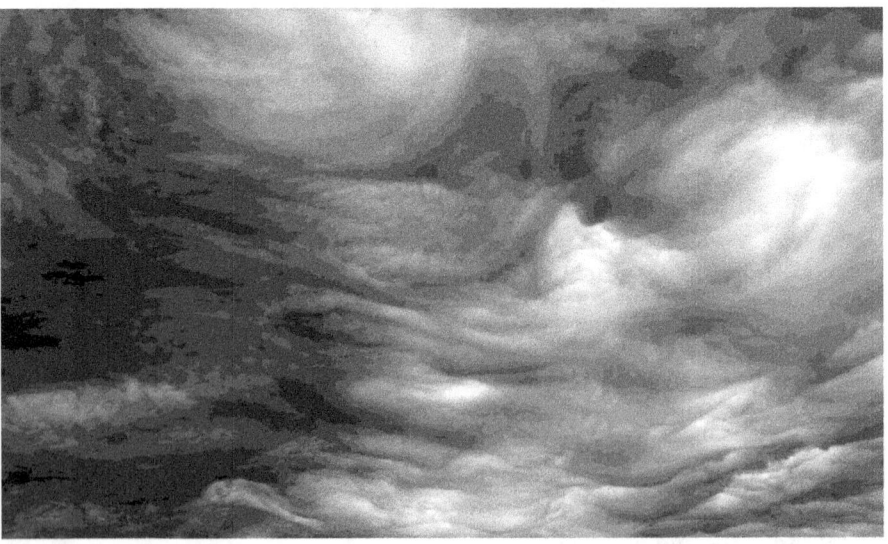

The presence of strange clouds, persistant haze and impaired visibility is not just in the air and sky, it MUST be having an effect on the ground, on the earth's soil and water.

The decline of the Natural World ... and Bioremediation

Francis Mangles, a retired federal wildlife biologist and botanist from Mount Shasta, California has been paying attention. He notes that between the 1950's and 1960's, jets did not leave many contrails. Today what appears to be jet contrails are much more frequent, and last for 10-20 hours. Here's what they look like.

"Counting all trails in every direction over Mount Shasta" Mangles Says *"FAA indicates that there can't be that many flights per day. The number has been as high as over 30 contrails at once"*, and I have counted more than one-hundred just before weather fronts. Typically heavy spraying days are just before weather fronts that come in from the pacific. So, here you have a man who is a retired federal wildlife biologist / botanist, and water specialist. Lab tests from northern California show very high levels of aluminum and barium in soil and pond water Francis says that *"1000ug per liter"*, when he was working for the federal government you were shut down. Sophia asked what is normal, what is tolerable? And he responded 0.5ug per liter is considered normal. Below are the findings from Mount Shasta:

Aluminum (ug/l)

0.5 microgram/liter considered "normal"

Mount Shasta, CA

Pond Water	**12,000 ug/l**
	24,000 x "normal"

Snowdrift at 8000 ft.	**61,100 ug/l**
	122,200 x "normal"

A lined pond (pond with a vinyl liner) in Bella Vista, CA measured a shocking amount of - Aluminum 375,000 ug/l, Barium 3,090 ug/l, and Strontium 345 ug/l.

Mangles has a master's degree in water related subjects, and he also reports that soil tested from outside a house in Northern California had 3,000 ug/l more of aluminum than from under the house. This means that something is dropping from above, and is affecting the soil around the house, but not underneath it. So you can't say that this is in the ground water, you can't say that this is in the earth itself coming from the earth. It's got to be coming from somewhere in the environment.

Tree Decline

In a book called *"The Dying of the Trees" by* Charles E. Little (1995), about the natural or biotic reasons for tree death, and also the effects of industry and de-forestation, which began decades ago, how trees, plants, and other life is dying from environmental pollution. Think of the manufacturing boom, acid rain, and pollution and how it was beginning to affect trees. But what has come along since Charles' book was written... is this

Now a days, we have an effect coming out of the environment that can't possibly be natural (wine-opener/corkscrew). You have nature trying to cope with the introduction of new substances particularly, metallic oxides which are also the metallic salts. These metallic salts have an affinity for water. Living things take up these salts along with the water that they are seeking. The natural world which gets its food from the ground is sucking this stuff up.

You can see how brittle the trees are now, lots of lost and /or brown needles, and if you look closely at the needles when you stand under many evergreens today, you can see that they are all frosted with brown. Signs of tree death are these sagging limbs, leafless scraggily branches, and infestation by bugs, insects, mites, fungus.

Many trees are now having to be felled due to severe mite infestations and other decay, that kills off the tree. All of the trees are dying. This isn't some sort of abstract or aesthetic problem, because the implications pose an existential threat. Without trees, all the species that depend upon them for food and habitat will perish. Even more crucial is that ultimately, and soon, crops will fail on a such a scale that widespread hunger will result. Scientists should stop putting their sole focus on the warming and weather destabilizing effects of CO2, and educate people about the effects of the "other" environmental factors instead, unless of course... this is what they want to happen.

If you investigate the bark on many evergreens, you will notice the bark becoming white on trees. You see white bark, and this white bark, when tested has had the following readings:

Solana Beach, CA
Tree Bark lab test

Aluminum 387 mg/kg
Barium   18.4 mg/kg
Strontium 113 mg/kg
Titanium   15.2 mg/kg

Mass analysis rather than liquid so the amount of materials is in milligrams.

Signs of tree decline; sagging branches, scorched look to leaves, secondary growth on trunk. When you see these growths ...

Singed, yellow and brown dying Oak and Maple leaves...
in Spring... not Fall

The beauty of an induced problem is that its twin is the solution.

Induced problem: Nature is dying

Solution: will come from science

As nature begins to die, we will be told that science has to step in to save it. GMO trees is the answer.

The Silent Forest
Genetically Engineered trees

Non-reproductive; No fruits, nuts, blossoms; no insects, animals, birds; low lignin (wood fiber), which will make them easy to cut and pulp.

The silent forest will grow straight and tall, and will be replenished by the state in what will be considered appropriate numbers, and in appropriate locations.

In May 2010, the following article states that the USDA approved large-scale field trials of 200,000 GMO eucalyptus trees made by ArborGen, a bio-tech company, are to be planted from Florida to Texas. We are told that the purpose of the trial planting is to evaluate whether such GM trees can become new sources of wood for paper and bio-fuels. This will also be done in the in the name of conservation and improvement.

May 2010 - GM trees from biotech company ArborGen (Article) "USDA Approves Genetically Modified Trees For Trial Planting"

The story they give is that they are trying to help, trying to "go green", trying to conserve, but then you find articles like this one.

2008: Scientists produce aluminum-resistant crops

"The plant is effectively blind to what's happening in the cell." - Paul Larsen, biochemist University of California Riverside.

MIT Technology Review: *A simple mutation to a single gene that makes plants thrive in spite of levels of aluminum that would normally be toxic.*

From the MIT technology review, an article announcing biochemistries' development of toxicity resistant crops, in particular in this article announces "*aluminum resistant plants*". So this article tells us that aluminum in soil stunts the growth of crops. Wheat, corn, and barley do not fare well in aluminum-laden soils, but now scientists have found a way to stop the plant from shutting-down its' own cell division. When a plant encounters toxins in the soil, it says to itself "I don't want to keep growing", so it shuts down the cell division. But they have figured out a way to keep prompting those plants to keep producing new cells. In other words this is a single gene mutation that activates a protein in the DNA of the plant, so that the plant continues to grow. The quote from this article is very interesting "the *plant is effectively blind to what's happening in the cell.*", from biochemist Paul Larsen. "*The aluminum laden plants can maintain high levels of growth in toxic levels of aluminum ... even if they sustain DNA damage.*"

When something begins to die in nature, it attracts bugs, molds; even viruses and bacteria. This is natures' way of hastening decomposition so that the dying form can become food for other living organisms. Today we have an epidemic of tree decline all over the world, not just here in the Pacific Northwest, or even just North America; this is happening in Australia, Europe, and many other (if not all) continents. In cities and suburbs, trees are rotting as they stand; swooning, and breaking. They are hazards to property requiring removal. I even have noticed some of the old growth trees around the courthouse in Washington County, Oregon and other areas have had to be removed recently.

Sunlight is a natural disinfectant, as hazy skies limit the sunlight, mold and fungus grow. As plants that are taking up toxin struggle to live. Molds, viruses, and bacteria begin to take them over. This is all part of nature, and bioremediation will be the obvious answer.

Bioremediation - Problem / Reaction / Solution

Metallic salts have made our air conductive. This means that we and everything around us can transmit and propagate energy. The air is no longer neutral, it no longer supports living things in a healthy way.

Fibers

The second group of materials found in these environmental samples, is "*unidentifiable fibers*", and I would really like you to appreciate the meaning of "unidentifiable" as it is used here. These fibers have been sent to sophisticated laboratories and there is nothing in the databases that match them.

Environmental samples show unidentifiable fibers ... *that do not exist in nature.*

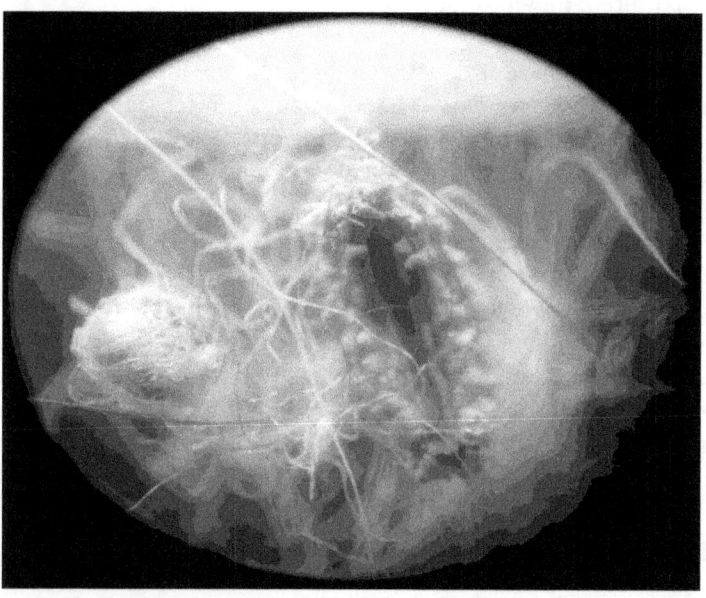

So , these are fibers that officially do not exist in nature. People around the world are developing legions on their skin that ooze and produce fibers. This is known as Morgellons syndrome. Tissue samples cultured from ordinary people without this ailment contain and grow the very same "unidentifiable" fibers.

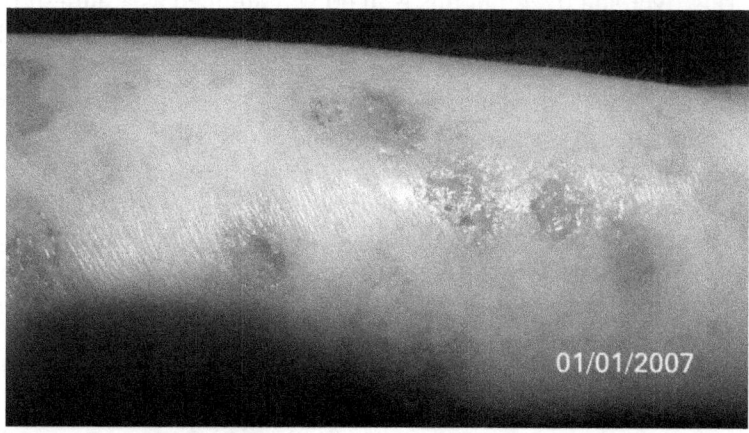

Lesion's have been reported to be larger than 5 inches in some cases. The fibers or filaments are actually tubules with hollow insides. When these fibers are cultured, they produce colonies of filaments.

These colonies continue to grow and reproduce, branching out into more filaments and more colonies. The filament cultures can be grown from; saliva samples, tissue from the skin, mucous, urine, and blood. In both animals and people, regardless of the presence of the Morgellons condition.

The fibers are segmented, with visible structures (devices) inside them. So where do these fibers and filaments come from? Airborne environmental samples that were collected by Clifford Carnicom (a researcher in this subject), in 2000 gathered at high altitude on a mountain in New Mexico, showed the presence of those fibers whose structure matched exactly the structure of those tubular filaments, also found in samples of blood, tissues and skin. The terms "fibers" and "filaments" are basically being used as interchangeable terms here.

Atmospheric fallout contains dried red blood cells

HEPA-filter samples contain embedded internal cellular structures (approx. 2 microns) similar to biological atmospheric samples. - "Extraordinary Biological Obsrvations" - Clifford Carnicom May, 2004.

Carnicom tells us that 3 main categories of a physical result are appearing over and over, but this is in no way implied as a limit of the types of material that appear. The physical size of these particles is at the sub-micron level. It is different whether you can see it or not, and these particles are not visible with the human eye. You need serious equipment to view particles that are in the range of 3 microns and smaller. A human hair, for example is 60-100 microns thick, these particles that we are discussing here are classified in the sub-micron range, which means that Even the physics of these particles are something that you are not going to be able to get your hands on. A standard HEPA filter that you can purchase for around $15 will capture particles all the way down to 3/10ths of a micron. These particles also have unique properties with respect to how they are affected by light as well.

The first of these sub-micron particle that is most commonly found in the HEPA filters is that of "ionizable metallic salts". So, what exactly is an "*ionizable metallic salt*" anyway? To answer this question, we must back up and examine what naturally occurring elements are common in nature.

Periodic Table of the Elements

H																	He
Li	Be											B	C	N	O	F	Ne
Na	Mg											Al	Si	P	S	Cl	Ar
K	Ca	Sc	Ti	V	Cr	Mn	Fe	Co	Ni	Cu	Zn	Ga	Ge	As	Se	Br	Kr
Rb	Sr	Y	Zr	Nb	Mo	Tc	Ru	Rh	Pd	Ag	Cd	In	Sn	Sb	Te	I	Xe
Cs	Ba	Lu	Hf	Ta	W	Re	Os	Ir	Pt	Au	Hg	Tl	Pb	Bi	Po	At	Rn
Fr	Ra	Lr	Rf	Db	Sg	Bh	Hs	Mt	Uun	Uuu	Uub		Uuq				

*Lanthanide series

La	Ce	Pr	Nd	Pm	Sm	Eu	Gd	Tb	Dy	Ho	Er	Tm	Yb
Ac	Th	Pa	U	Np	Pu	Am	Cm	Bk	Cf	Es	Fm	Md	No

* *Actinide series

If we look at the periodic table of the elements, one might notice that many of the elements are metals. We could even expect sub-micron size particles of these elements in the atmosphere, but it becomes less likely.

Then these elements must be oxidized or made into something that can be soluble, or a metallic salt, which is even less likely to occur without mans assistance in nature. If you know much about chemistry, you will know that adding even the slightest bit of salt to water changes the entire electro-chemical characteristics of the water. This change makes it so that the new salt-water solution will now carry a current. Pure water carries very little current, almost nothing; but just add a little salt and everything changes as far as the physics of that solution.

Now that we know a little more about what a metallic salt is, what separates these particles once again from being naturally occurring elements is that these metallic salts are also ionized. When a particle is ionized, all that this means is that it is charged. So these particles are electrically charged, soluble metals that are so small that you need a pretty sophisticated microscope to even begin to view them. This also means that these particles have the ability to conduct a current.

The ionosphere is a layer up above us, it has very different physical properties, and if you look at the amount of the ionosphere, 2-3% of that layer is electrically charged which completely changes the physical properties of anything in this environment. So when it is said that we have the repeated detection of these sub-micron sized *"ionizable metallic salt"* particles; the expected consequence of that is that our atmosphere, which by all means is not regarded as a conductive environment, or at least not highly conductive in any way, is *"enhanced"* by the propagation of electric energy through the environment.

Air is generally not conductive, and we know this because it is one of the best insulators there is, but what if the physical properties of the environment are changed by the introduction of soluble metals that have the ability to carry a charge? As a result, the basic physics of the entire planet are changed. There is a shell around the earth, and the term plasma comes up, and as it turns out it is estimated that 99% of the universe is made up of it. All that plasma means is *"electrically charged gas"*. So what happens if you consider the shell of our earth (the ionosphere), the very physics of that being changed in an electromagnetic way, not just in a physical way. If we take it from the simplest level, these particles being released into the atmosphere are at their most basic level, simply pollution.

At a very fundamental level, if these planes cannot keep from generating pollution, we have the responsibility to regulate the pollution that is released into our environment, but I really fear that there is something much more sinister going on. We need to look at it from the view of human life; the more people that inhale these particles, in the simplest terms more people will die from it. So on the very basic physical level we need to view this as a hazard and risk to our health, but the ramifications of what I am acknowledging here are much, much deeper in terms of what the impact is on not only human, animal, and plant life, but on the entire planet as a whole.

This brings us to the next most common material found being released into the environment via aerosols, and these are the fibrous materials. As noted before, these *"unidentifiable fibers"* do not occur naturally at all. These materials are also showing up over and over and over, and they are also sub-micron in size. Normal fibers are on the order of 7-12 or so microns thick, you can see them with your eye. Think of wool or cotton fibers... they are thin, but you can see them. Another example is a spider web, this is an incredibly strong material, and is also only around 7 microns thick. An asbestos fiber is around 2 microns... still not as small as these fibrous materials falling on us from the sky. Yet I am sure that no one would want to make a habit of breathing asbestos fibers. Many of us are familiar with the environmental attention that has been given to the asbestos issue in our lives; what kind of issues and problems could we expect with these "*nanofibers*"? What if you have a material in the atmosphere that is measuring at the sub-micron level? This particular material was actually even sent to the United Stated Environmental Protection Agency, with a very reasonable letter sent by certified mail requesting identification of exactly what this material was, on behalf of the public interest.

If you look at the mission statement of the EPA, it should be perfectly reasonable to ask them to identify a material that you believe may be hazardous to living organisms. The EPA refused to acknowledge even the physical receipt of that material, there was correspondence, they did write back, but with no reference at all to the receipt of the physical material. In the response from the EPA, a few choice words were used over and over which were "*We are not aware*". "*We are not aware of any aerosol operations.*" The EPA also stated that it was not the policy of the EPA to identify any "*unsolicited*" material.

The third most common material found being released into the environment via aerosols, are the "*biological components*", also known as desiccated (freeze dried) erythrocytes (red blood cells). Essentially every one of us, the vast majority of us appear to *already* be harboring and carrying a set of pathogenic forms that to the degree that I am able to determine; has a direct tie to the aerosol operations. I think that the biggest question that we need to be asking is why are these dried red blood cells in the air? At the very least, this is extremely puzzling.

A medical microscopist or biologist that specializes in microscopy has confirmed the existence of these dried red blood cells, and went even further and confirmed that they are actually human red blood cells, and that they had been engineered in some way as to be preserved. So, again, we must ask ourselves... if I were to approach this as a child and were to say "*Mommy, what are red blood cells doing falling out of the sky?*"

The Three Kingdoms of Biological Life

Biology defines life into three kingdoms, each with different organisms classified into groups based on their characteristics. All biological life on our planet can be classified into these three kingdoms or domains; these domains are Bacteria, Archaea, and Eukaryotes. The Archaea are the heartiest of the life forms; these simple organisms have no cell compartments and are able to tolerate extreme conditions (such as grinding pressure, extreme temperatures, or extreme acidity or alkalinity) that most other organisms would find intolerable. They can live in volcanoes, geysers, and on the ice shelf. Bacteria, on the other hand will die when exposed to extreme heat or cold; this is why we cook and freeze our food. Bacteria are also simple organisms without internal cell compartments, yet they are unable to tolerate the extreme conditions that Archaea can survive in. The third category are the Eukarya (which is what we are); and this term refers to complex organisms with defined cell compartments and internal cellular components.

Bacteria and Archaea are simple organisms with no sub-cellular compartments. Eukaryote are complex with defined cell compartments, and many internal organs (such as mitochondria that make DNA and provide energy to the cell). Plants, animals, and humans are Eukaryotic; as are fungi and slime molds.

You can get a better idea of which life forms are in which category with the "Phylogentic Tree" below.

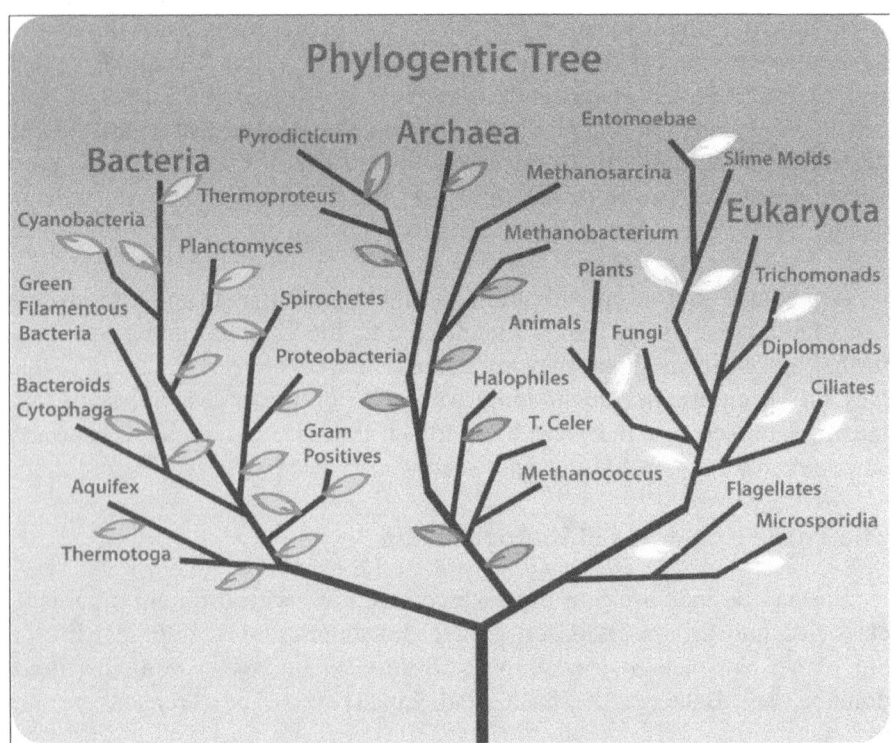

Now the materials appearing within the fibers and filaments mentioned previously are as tough as archaea, they look like bacteria, and they self-replicate.

One looks like something called mycoplasma, but is not. One looks like Chlamydia, but is not.

Natural fibers or filaments would be classified as fungal in the domain of Eukaryotes, but these fibers contain forms from the other two groups inside them. The next image is of a tubular fiber at 5600x magnification, with self-replicating internal elements, resembling bacteria and behaving like Archaea. This DOES NOT HAPPEN in the natural world. So what exactly are these fibers?

What is going on here?

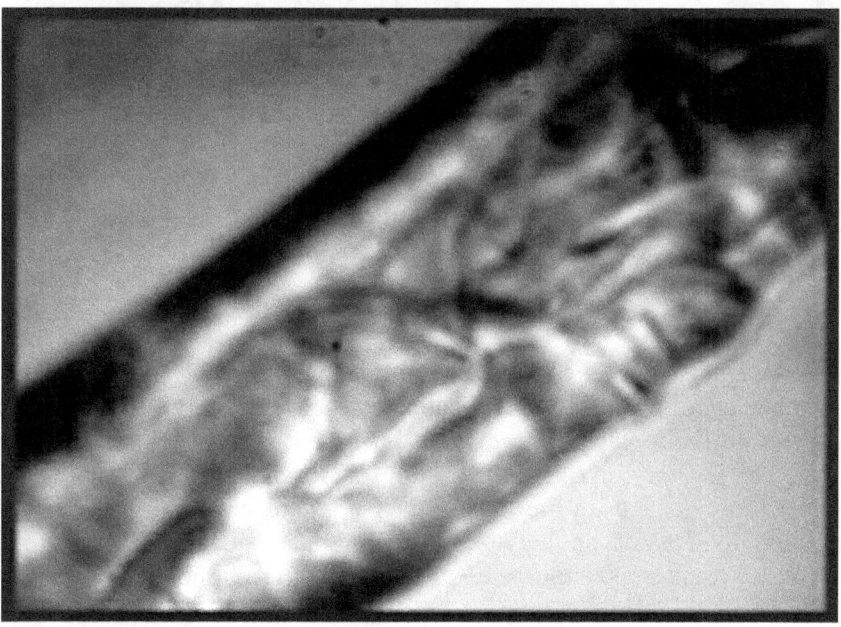

Tubular fiber (5600x magnification) produces hybrid life forms. Filament with internal self-replicating elements resembling bacteria and behaving like Archaea.

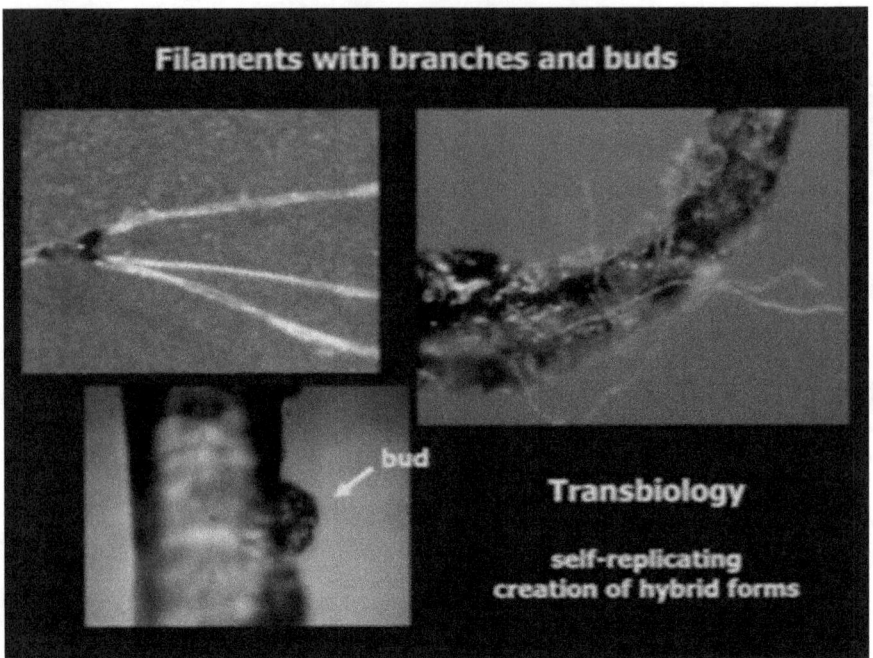

It appears that within our bodies we now have all three life kingdoms, replicating themselves... what does that make us? Are we still Eukaryotes, or are we becoming something else? You can call this "trans-biology", or the crossing of biology. The creation of hybrid forms. Materials are forming in our bodies currently that are not native to us, not natural, and entirely new.

You can see in the previous image the tubular filaments with the branches shooting off (self-replication), creating more colonies. You can also see in the lower portion a little bud. That is the beginning of a new fiber. The territory on the scale of atoms and molecules, is the nano-world. Science has opened up the nano-universe where incredible new creations are possible. Nanotechnology explores materials that are less than a micron in size, or from 1-100 nanometers. A nanometer is a billionth of a meter.

Nature and biology continually work the nano-scale, assembling proteins and building with crystals to conduct the business of life. Spider webs are an example of very fine filaments with enormous strength and flexibility because they contain nano-scale crystals.

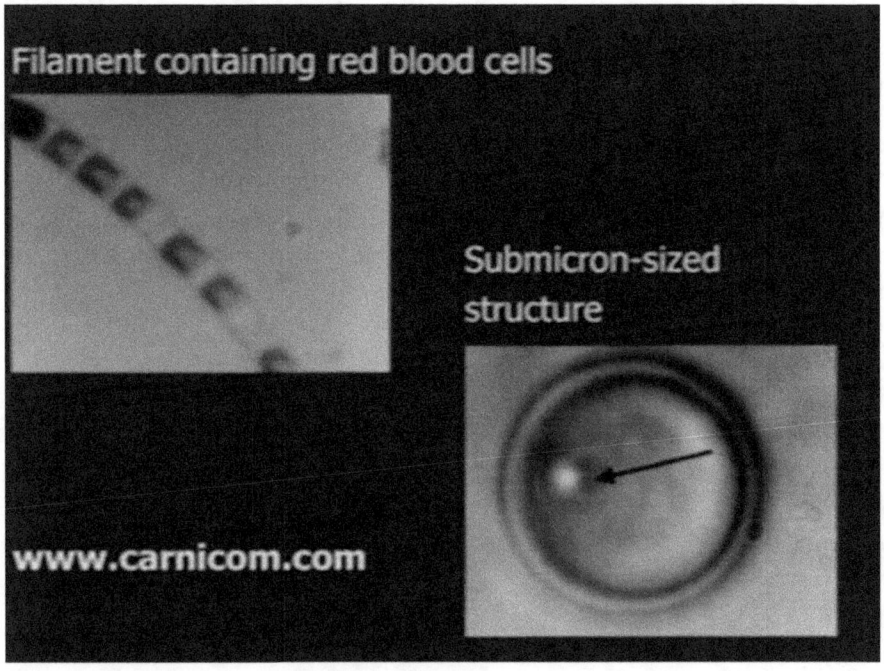

Let's go back to the filaments and the structures that they contain, these ones that Clifford Carnicom has found. In those unidentifiable filaments, he has observed the formation of red blood cells and submicron-sized structures. So now we have a filament, making its own red blood cells.

The engineered red blood cells are very tough, withstanding excesses of heat and chemicals, indicating that they are designed to endure almost anything.

He has put them in a Bunsen flame, poured bleach on them, acid, and they still endure. In addition they are able to replicate, growing outside of the body in a Petri dish.

This is highly sophisticated technology, going on by itself, not even in a laboratory. So we are witnessing red blood growing outside the body, having nothing to do with bone marrow, in a Petri dish in a lab.

Could it be, that artificial materials are being introduced into living things?

Here you have what Sophia Smallstrom refers to as the "Nanotech-Pyramid", which comes from the nanotech industry. You start with materials, which in turn, build and form structures, which in turn create processes with which devices are made, and as a result complex systems can then be created.

NANOTECH PYRAMID

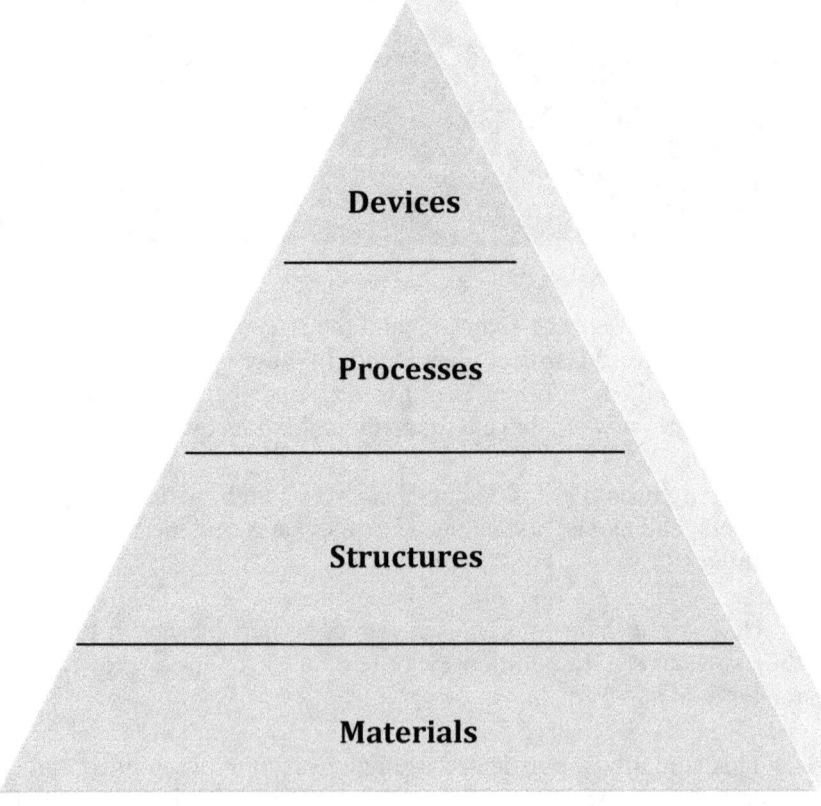

Devices

Processes

Structures

Materials

Could it be that artificial materials are being introduced into us to make new processes from the ground up? And will these processes override our own natural biology? Our own internal systems?

People with advanced Morgellons syndrome, started with fibers coming out of their skin, and they are now observing very strange crystalline forms, and metallic devices. The fibers became multicolored, and now they were accompanied by plaques, strange hard little things that resembled shards of colored glass. The lesions keep emitting these things, and do not heal over or close, sometimes for years.

Below are more images of plaque structures, and metallic grooved devices.

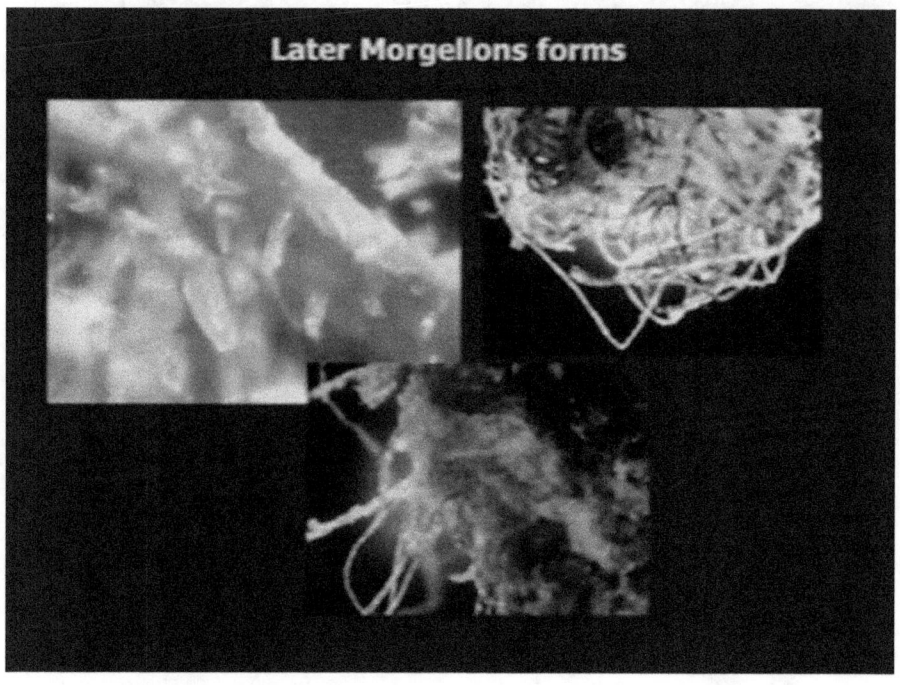

The fibers became multicolored, and now they were accompanied by plaques, strange hard little things that resembled shards of colored glass. The lesions keep emitting these things, and do not heal

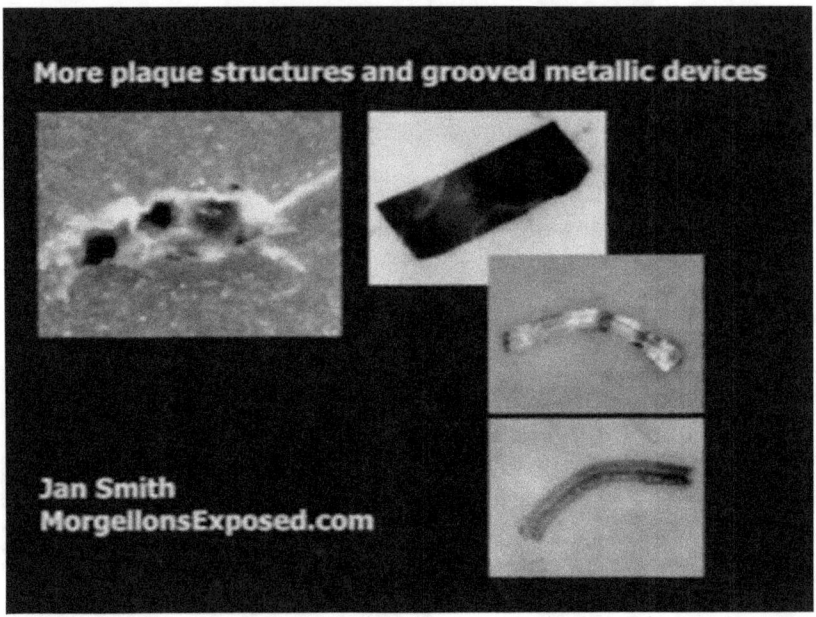

In this image, on the right, is the front and back of a metallic tubular device found inside the skin.

These images are provided by Jan Smith of MorgellonsExposed.com who has done extensive research on biotech websites looking for things that match what comes out of her body, and she is finding them.

The fibers from her skin, were from high-density polyethylene, and on her website she says; "*The bizarre nature of my findings suggest a man-made source. It occurred to me that if these pathogens were being bioengineered in a lab, they were made of multiple components. The mutated material might reproduce, and intermittently send out a batch of identifiable debris much like the original genetics.*" These type of debris are also called "throwbacks", and perhaps we can start to gain a better understanding of exactly what is happening by examining these materials because they more closely resemble the original pathogen, and not it's product/s.

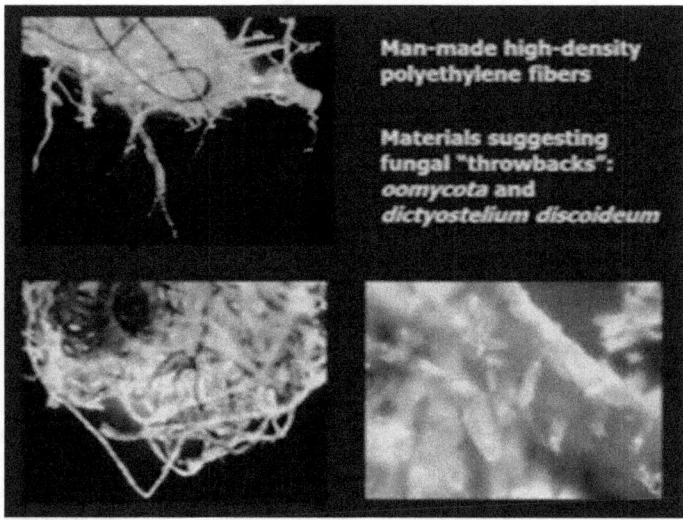

It is Jan's theory, that when new life-forms are made through this gene splicing and mutation, bioengineering and nanotechnology; they will self-replicate in all sorts of ways, creating different mutations in each generation.

Morgellons specimens can sometimes be identified with the original spores, before those mutations occur. Jan found three varieties of something called Oomycota fungus in her specimens, as well as something else called *Dictyostelium* DISCOIDEUM, which is the slime mold that is a major player in bio-medical research today. The material coming from her body was ribbon-like and motile, it was grooved and oval very much like the Oomycota you see below.

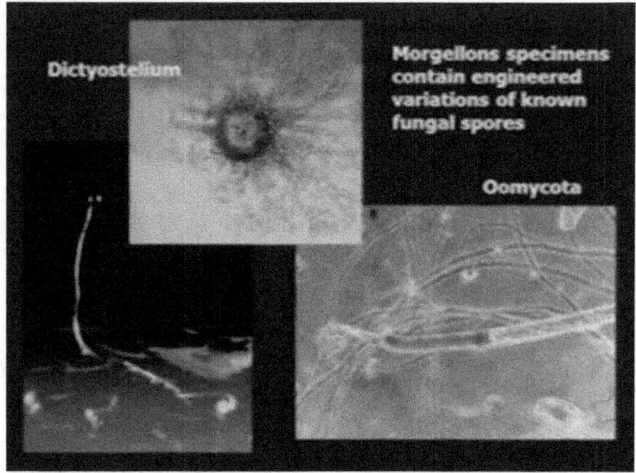

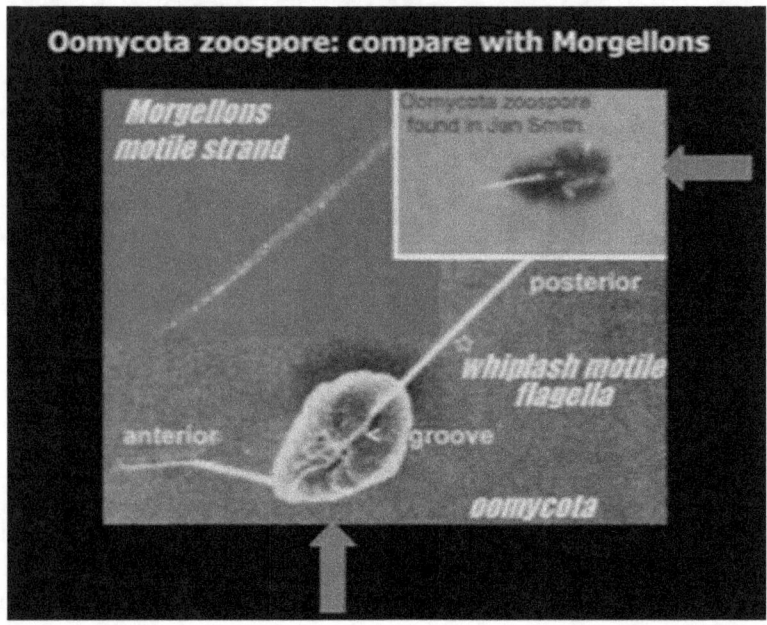

The arrow on the bottom, points to a diagram of Oomycota; and on the upper right, you see something that came out of Jan's body that looks exactly like Oomycota. The fiber strands were made of cellulose and GNA (Glucanspoly Saccharides, a type of sugar), and was confirmed by microscopy. GNA is a synthetic cousin of DNA, it has a three carbon structure versus the five carbon structure of DNA.

"...the first self-assembled nanostructures composed entirely of glycerol nucleic acid (GNA) - a synthetic analog of DNA. The nanostructures contain additional properties not found in DNA, including the ability to form mirror image structures." From the article: **GNA: DNA's Chemical Cousin is a Nanotechnology Building Block** - Science20.com April 28, 2008.

A NIH article from Japan, on research from Japan tells us that various conductive polymers and gold nanoparticles are entrapped within the helical super-structure of this GNA. The particular sugar found in this GNA is called Beta 1-3 glucans.

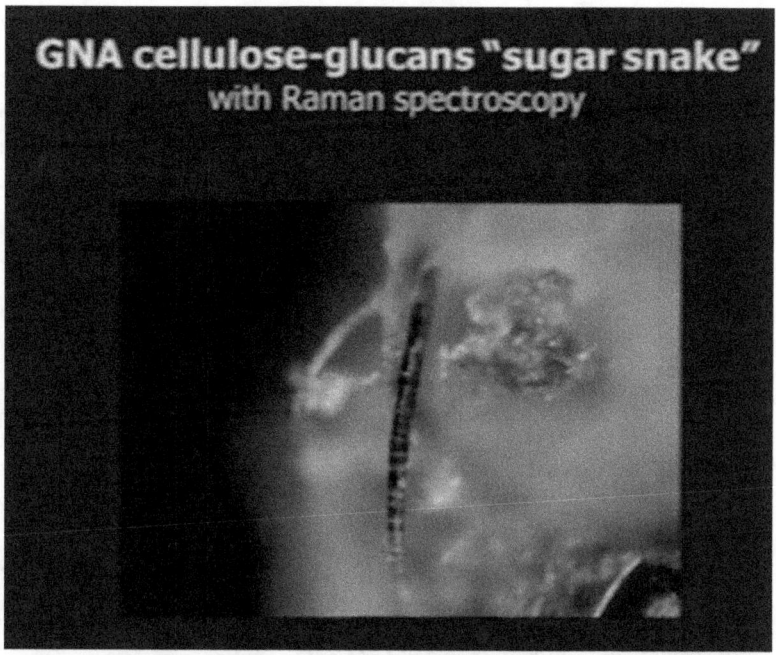

The above image was captured with a Ramen Spectroscope, very sophisticated equipment. The microscopist who looked at this called it a "*sugar snake*".

When Jan's fibers were put to a high-heat flame, they released a what she called a "*gold payload*".

They would drop out this tiny little bead of gold. Pictured below are more of these gold payloads.

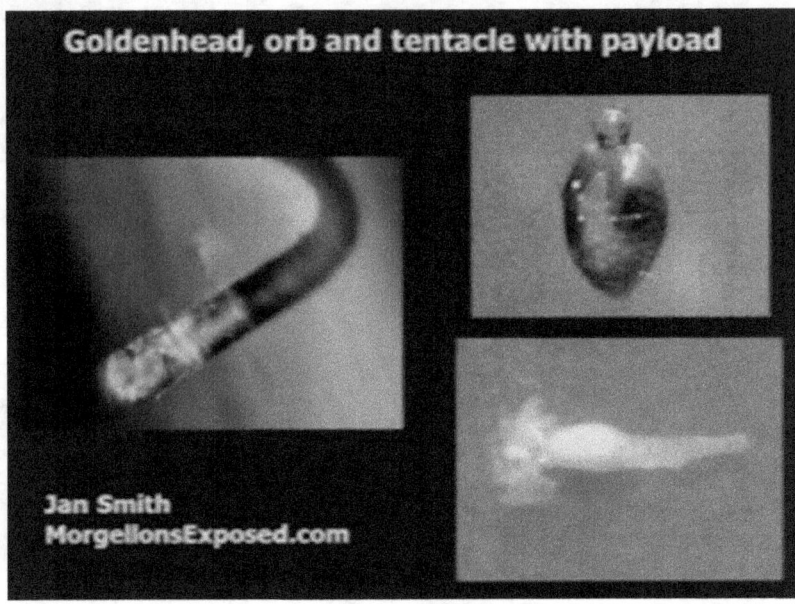

These fibers would contain a "head", of which Jan would call them "*goldenhead*" orbs and they would all have this payload in them.

Jan Smith also found industry matches to a device called a "*nano-array*" in samples taken from her body.

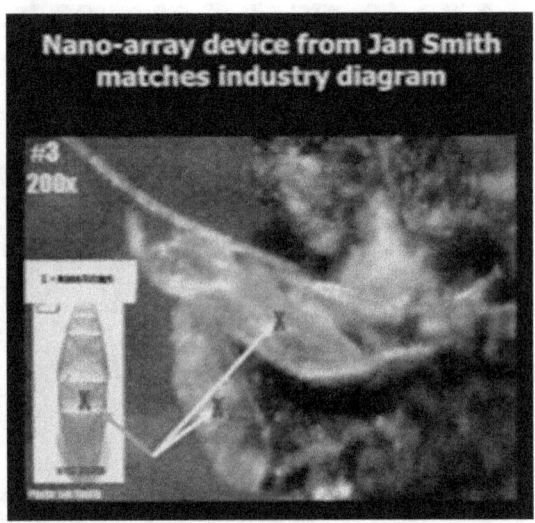

A "*nano-array*" is a tiny device used in biotech for DNA hybridization, in the image above; on the left is a diagram from an industry website of a nano-array, and in the image above there are actually two of them marked with "X"s. These came out of her body, and looking exactly like this.

Why exactly would someone be finding devices in their body that are used for DNY hybridization?

Remember the nanotech pyramid in an earlier diagram? We begin with the emergence of basic filaments, followed by more complex structures. What processes are going on here? How are these materials combining to form devices that are working together inside them.

What is happening to our biology?

Remember that tissue samples obtained from ordinary people who have no Morgellons symptoms or lesions can be cultured to produce the very same filaments found in people with Morgellons. We would have to conclude that Morgellons is like the canary in the coal mine, and that only some people are exuding the materials. Could this be because their bodies are rejecting it, while our bodies are integrating it?

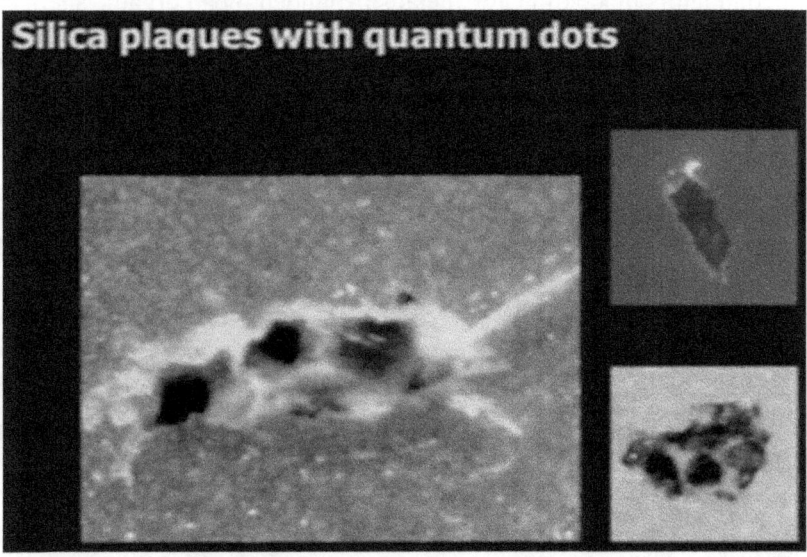

Jan's body has been sending out these colorful plaques, which are hard pieces of silica. Some of them have dots on them. They are very small, they require handling with a needle as they are placed under a microscope. The plaques are fragile, they can shatter.

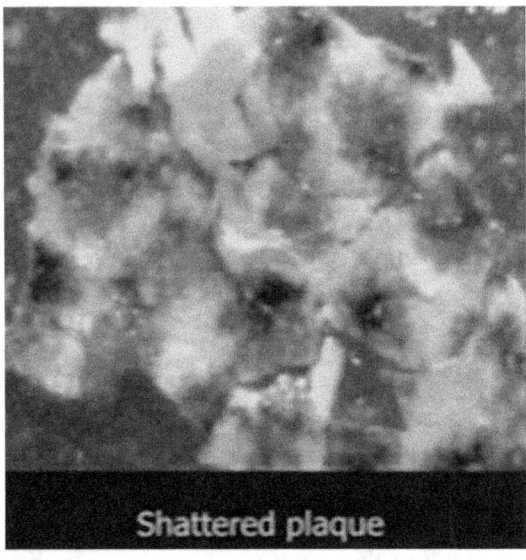

Quantum dots which are colors visible in the plaques, are nano-crystal semiconductors made of heavy metal surrounded by an organic shell.

Jewel-like hexagons, faceted pyramids, crystals. Hexagons have been found environmentally, not only in tissues, but they are in environmental fallout. The specimen below is provided by Morgellons Research Group.

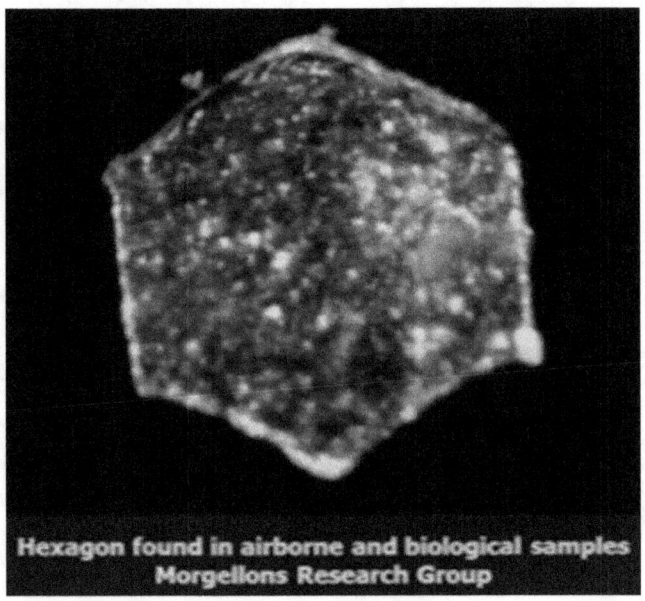

Pictured below is a pink almost ruby-colored, layered crystal. Visible in the smaller inset image are the layers. This crystal was found in the environment.

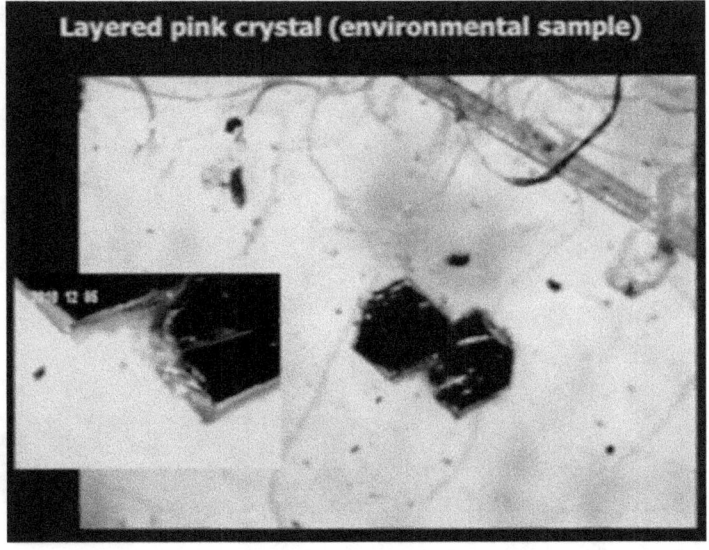

The contents of the image below are not visible to the naked eye, but this image is from a woman that lives in Oregon that was examining a hay stalk with a microscope.

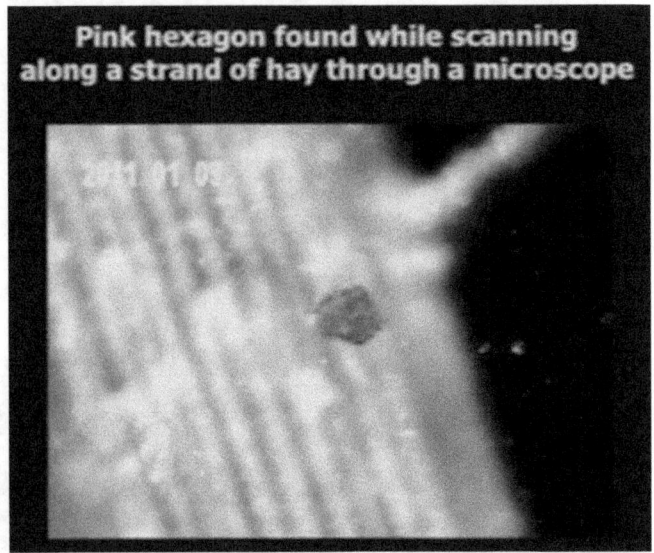

There have also been samples found with embedded hexagons and also strands that glitter (pictured below right, these samples came from a living human body).

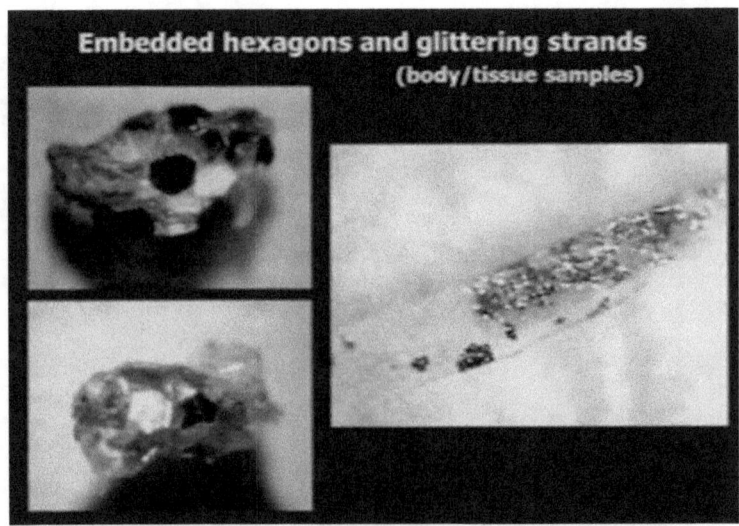

In the image below of some material removed from a lesion, notice the fiber coming out of the bottom, and the crystals apparently growing from the living tissue.

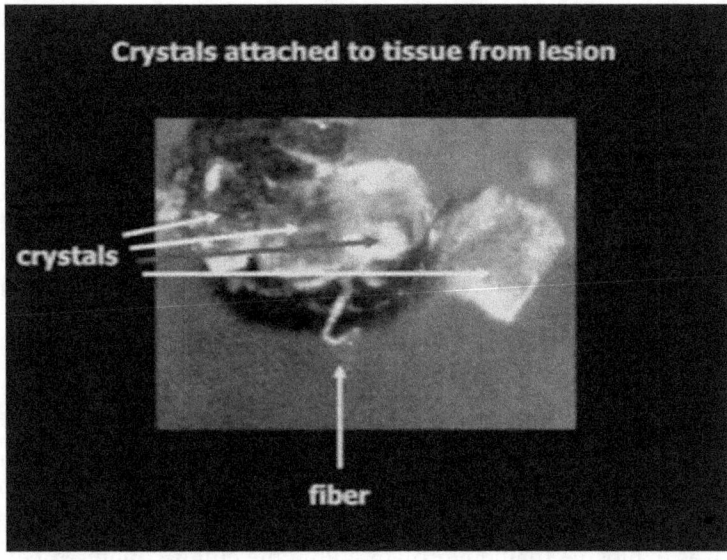

There are also combinations of wires and multi-colored quantum dots (pictured below).

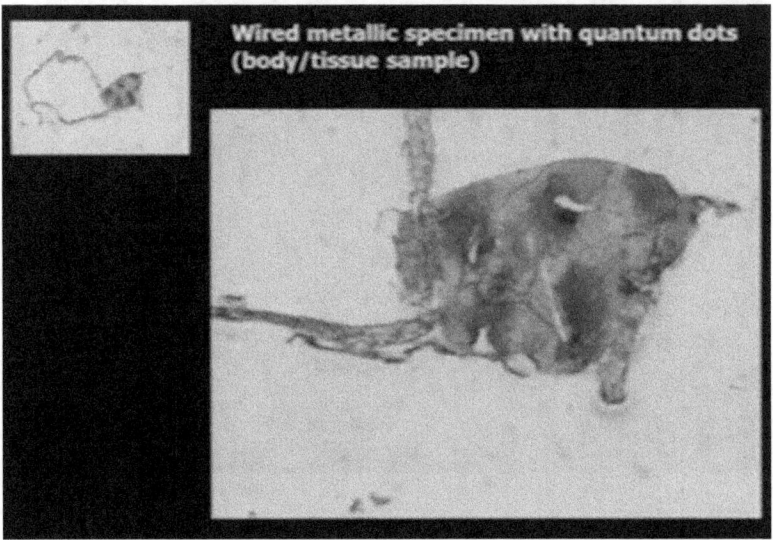

The industry tells us that quantum dots are tiny nano-crystals whose small size gives them unprecedented *tunability.*

Tunability? Also known as the piezoelectric effect; is the internal generation of a voltage, when an outside pressure is applied and vice-versa.

The piezo-effect occurs in crystals, ceramics, DNA and certain proteins. An example of this would putting a frequency to a crystal, it would generate a voltage; and when you put a voltage to a crystal, it responds with a frequency. We call this Piezo Electricity, which is electricity resulting from applied compression or frequency. A common everyday use of this technology is found in many electronic lighters, as a matter of fact, enough electricity is generated by the downforce applied to the piezoelectric material that not only can it cause a spark to light the lighter but also even power several LEDs at the same time until they fade as the charge is drained.

Recent work by Clifford Carnicom reveals that a filament culture, subjected to a blue light frequency (375 nm) resulted in explosive growth of the culture within 24 hours after the initial incubation. In other words, he put the blue light on, and then incubated the culture for 5-7 days in which time a film developed over the culture, and then within 24 hours after that there was explosive filament growth within the culture (pictured below).

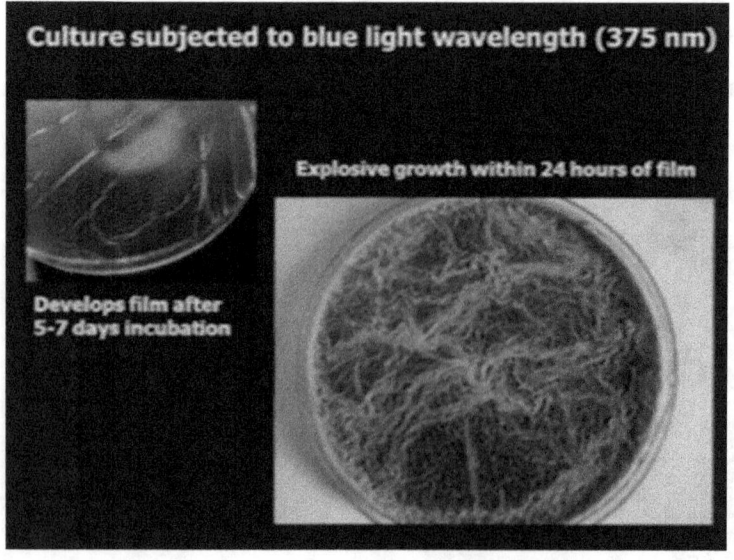

Culture subjected to blue light wavelength (375 nm)

Explosive growth within 24 hours of film

Develops film after 5-7 days incubation

We now live in a frequency-filled world.

HAARP, Artificial Weather, and there are an enormous spread of unnatural frequencies that affect the earth today.

I came across a very well known paper about electromagnetic fields and neurology (specifically neurological function) written by Prof. Ross Adey at Loma Linda School of Medicine, San Bernardino California; with a subtitle *"A Possible Paradigm Shift in Biology"*. In which we learn that the earth's natural frequency peaks at 32 Hz, and this high of a natural frequency only occurs in equatorial thunderstorms.

The electronics we use today generate electromagnetic fields that are tremendous in the MegaHerts and GigaHertz ranges (Million and Billion hertz, respectively). In the image below, you can see earth and nature at the top with an extremely low frequency of 3 Hz - 30 Hz at the 100,000 km - 10,000 km wavelength. Then we see that HAARP has a low range of 30 Hz - 300 Hz, then the "Ground Wave Emergency Network" (or GWEN) has a range in the low KHz range, then more HAARP in the low MHz range, another GWEN in the mid MHz range, then in the GHz range we have a plethora of mobile and entertainment frequency generating devices.

ELECTROMAGNETIC FREQUENCIES

Designation		Frequency	Wavelength
			Earth/Nature
ELF	extremely low frequency	3 Hz - 30 Hz	100,000 km - 10,000 km
SLF	super low frequency	HAARP 30 Hz - 300 Hz	10,000 km - 1000 km
ULF	ultra low frequency	300 Hz - 3000 Hz	1000 km - 100 km
VLF	very low frequency	GWEN 3 kHz - 30 kHz	100 km - 10 km
LF	low frequency	30 kHz - 300 kHz	10 km - 1 km
MF	medium frequency	300 kHz - 3000 kHz	1 km - 100 m
HF	high frequency	HAARP 3 MHz - 30 MHz	100 m - 10 m
VHF	very high frequency	GWEN 30 MHz - 300 MHz	10 m - 1 m
UHF	ultra high frequency	300 MHz - 3000 MHz phones	1 m - 10 cm
SHF	super high frequency	3 GHz - 30 GHz & internet	10 cm - 1 cm
EHF	extremely high frequency	30 GHz - 300 GHz	1 cm - 1 mm

Professor Adey reminds us that chemical bonds are magnetic bonds, formed between atoms by paired electrons with opposite spins which are attracted magnetically. So if nature itself moves in the 3 Hz - 30 Hz range, what is happening to us on the biochemical level with all the different frequencies that our bodies are exposed to and experiencing? What is being done to our biology?

Again, I will repeat Ross Adey's tenet; *"Chemical bonds are magnetic bonds"*.

Electromagnitism is capable of changing what is happening in our bodies.

Frequencies are capable of supplying to the synthetic materials in our bodies, the force or power that activates them gets them working, and makes them come alive.

Nothing about us today is normal; we have hair that glows, and skin that shimmers. The image below is from a healthy woman in Oregon that has been doing microscopy (iridescent effect) .

In another instance a glowing metallic object was found while scanning rainwater under a microscope (below).

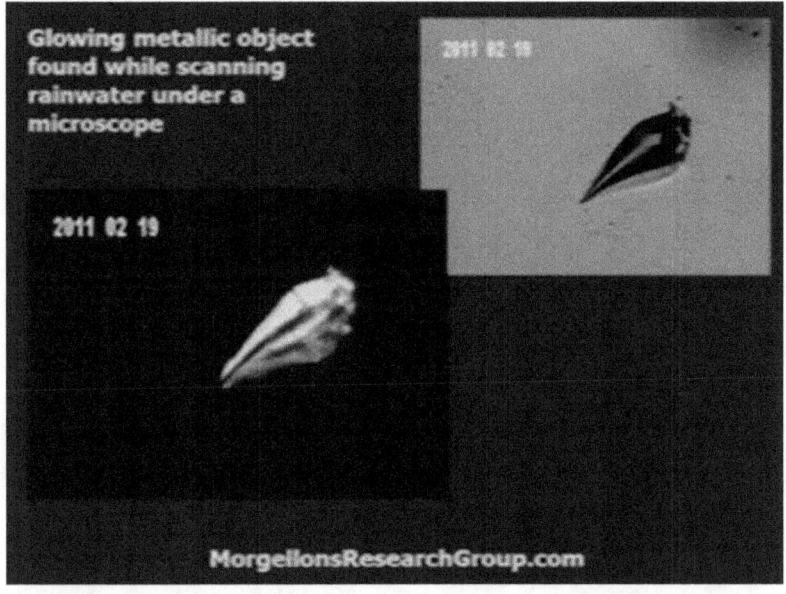

And in yet another image we find an instance where a hexagonal crystal grew a fiber in 20 seconds while being observed under the microscope.

The field of synthetic biology is a new and still emerging frontier of science. It draws from Biochemistry, and Biomedicine, Genetics, Robotics, Radiation Biology, and Information Technology. Using nanotechnology, it's goals are to improve and transcend the limits of nature.

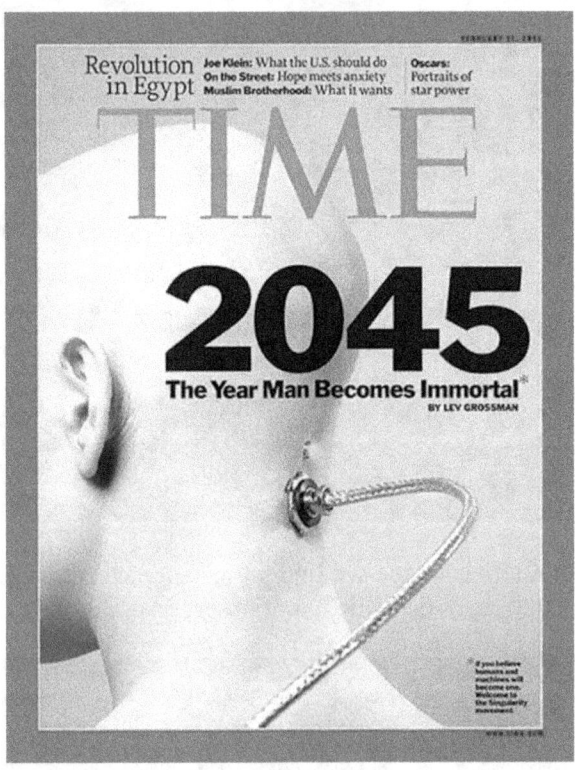

February 21, 2011; time magazine had a feature story about the Singularity, which is a term that signifies the merging of man with machine. The story was released on February 14th, Valentine's Day; to underscore a love affair... the marriage of humanity and technology.

The article tells us that a transformation is coming, and our species Homo Sapien will no longer be recognizable as itself.

We will be something new, something better. The time predicted for this transformation is 2045.

The man who has made this prediction is Ray Kurzweil; a futurist known for his uncanny accuracy in precisely this area. The pace at which technology grows and improves, such that it will one day be smarter and better than us. This is the Singularity.

There is a singularity university hosted by NASA, and sponsored by Google. This university purpose is to teach people about the intelligence explosion. Ray Kurzweil has made fortunes over and over as an engineer and inventor. A documentary about him is called "The Transcendent Man". He wrote the bestselling book "The Singularity is Near", released in 2005. Singularity is a word from astrophysics, referring to a point in space-time where the rules of ordinary physics no longer apply.

Kurzweil has correctly predicted the growth of information technologies. He has made it clear to the world that technological progress is exponential, not linear; which means that advancement begins to advance itself, in a manner of speaking. Exponential curves start slowly, and then explode.

A quote from the Time Magazine article mentioned above; "*In Kurzweil's future, biotechnology and nanotechnology give us the power to manipulate our bodies and the world around us ... at will, at the molecular level. We ditch Darwin and take charge of our own evolution.*"

The question is ... who is 'we'? And what is 'at will'? *Whose* will?

Kurzweil predicts that by 2020 we will have successfully reverse engineered the human brain, and when hyper-intelligent, artificial intelligence arrives; all we have to do is hand ourselves off to it. Armed with advanced nanotechnology *A.I.* will solve the problems of the world.

Strong A.I., is a super-powerful, broad-spectrum intelligence that operates as easily and comfortably as a natural (or wild) human being. It is by no means simply a chess-playing computer we are talking about, it is a machine intelligence that can pass for human in a blind test. This is as close as you can get to consciousness or sentience, which is the ability to feel, perceive or be conscious, or to have subjective experiences.

When this *'Strong A.I.'* finally arrives, what will it do as a newly created inhabitant of the Earth? Would it compete with us for resources? More intelligent than we are ... would it treat us as 'lesser' beings? Would it even recognize that we are the ones that made it? Or would it overrun us?

Kurzweil is one of the world's leading transhumanists, actually number 30 on Times 'Most Influential' list. Transhumanists believe that we, ourselves *should* merge with machines. Imagine a time when we can download our brains into a computer, and/or upload a computer into our brain.

"The three great overlapping revolutions ... sometimes go by the term **'GNR'**. 'G', stands for *Genetics*, or biotechnology; is mastering the information processes in our biology. We actually eventually will be able to reprogram biology away from disease and from aging. 'N', stands for *Nanotechnology*. In the next 25 years we *will* have blood cell-sized devices that go inside your body, and keep you healthy from inside. That go into your brain and interact with your biological neurons and allow us to merge with non-biological intelligence. The third one goes by the letter 'R', which stands for *Robotics* or robots. Really though... we are first to artificial intelligence, and that's the most significant revolution of all. In about 20 years ... I have set the date 2029 ... a machine, an A.I. will be able to match human intelligence, and go beyond it. Artificial Intelligence, which will give us more than human intelligence, but will actually be able to give us super-human intelligence. Will allow us to solve problems that we are unable to solve today." - Ray Kurzweil, from *"The Transcendent Man"*

Once again we hear mention of artificial blood cells in "*The Transcendent Man*".

"*Biology is very impressive, intricate, clever, but also very suboptimal compared to what we will be ultimately able to engineer with nanotechnology. We are building devices now that are at the nano-scale. This is the design for a robotic red blood cell. Conservative analysis of these red (rescrothites), if you would replace a portion of your red blood cells with these robotic versions; you could do an Olympic sprint for 15 minutes without taking a breath, or sit at the bottom of your pool for four hours. We will be able to download software against specific pathogens that have never been seen before, subject autoimmune disorders and if you look at what will be in principle with nanotechnology we will be able to go far beyond the limitations of our 'version one' bodies.*" - Ray Kurzweil, from "*The Transcendent Man*"

The robotic red blood cells will supposedly improve us and keep us healthy ... as he says. What if this is just the 'gloss' to sell us on such bodily intrusions. Just as we are being sold on the idea of a 'smart' environment; a techno-matrix that will vibrate with not just intelligence, but also connection.

What indeed, is the 'world wide web'? We may think that it is the 'internet', but it will be the humming network of everything connected through ubiquitous intelligence, or intelligence that is 'everywhere'. Artificial Intelligence will connect the world; Homo Sapien will be transformed into '*Homo Evolutis*'. Biological processes will be run by technology. Living things will not be reproductive; the world will be populated with engineered species, and all processes will be patented, licensed, and controlled.

You could consider nanotechnology; the installation of artificial intelligence, in living and nonliving things. Smart dust for example are tiny nano-sensors that can float and land anywhere. As Kurzweil declares "*Self-replicating nano-technology will infuse everything around us... with itself.*"

"*As a person I am human, and I am really limited and restricted in what I do, so if I could come out of the singularity being mentally and physically upgraded... yeah I would go for that. I don't mind changing dramatically from what I am.*"

"I believe fully that there will be flash memories, that you can plug into your brain. We'll be able to hook our brains into calculators and statistics programs... have Google directly into the frontal lobe. There is going to be a lot of expansion of the mind through interfacing the human brain with technology. There is an unanswered question, of how far can you go, and still be human?"

"As we merge with machines, and I think it's inevitable that we will; we will transform into something new."

"And as the technology becomes vastly superior to what we are, then the small proportion that is still human, gets smaller and smaller and smaller until it is just negligible."

"Anybody who is going to be resisting this progress forward is going to be resisting evolution, and fundamentally they will die out. It's not a matter of whether it is good or bad, it's going to happen."

You have seen today, the deposition and active presence of artificial materials in the sky, environment, and living things. Nanotechnology has arrived at our personal doorstep without our permission. It isn't that this will happen in 2045... it is already here. Human enhancement is being sold to us as the ability to become super-human. Having better, higher, faster intelligence, perfect health... all of this is a sales pitch. Enhancement may in fact be degradation, or actually being devolved to someone else's specifications.

What is ... enhancement ... improvement ... better performance?

While nano-biotechnology promises in headlines to make our world better it may in fact be busy taking us over; so it can tailor us to the plan for *'the hive'*. Already, transhumanists are looking forward to the creation of the post-human, an improved human that will have no gender, will not reproduce, will be a better *performer* in the workplace, will not be distracted by love or lust, will be free of disease (thanks to nano-bots keeping every one healthy); but all this is part of the fantasy.

In reality, thanks to stressors on our physiology, infertility is already soaring, our sexuality is diversifying, and the nuclear family is falling apart.

Biotech is an exploding frontier; it is cleaver enough and small enough to enter and change our very cells. New forms of DNA have been invented;

there is GNA as previously mentioned, as well as PNA (a hybrid of protein and DNA), that will add to our double helix... a third strand.

When nanobiotech has a firm footing in us; it will be easy to 'upgrade' and 'downgrade'. Anyone and anything; in any way.

Oliver Curry, an evolutionist at London School of Economics predicted in 2007; "*The human race will one day, split into two separate species.*" An attractive intelligent ruling elite, and an underclass of dim-wited, ugly, goblin-like creatures.

So here you have the *'e-workers'* and the *'e-lites'*. Transhumans will presumably be involved in this process of transformation, the process of 'renovation', or remaking us into what someone considers improved.

We are Transhumans *now*.

Improved is only what fits certain specifications. For instance; a specimen that can work 18 hours-a-day, a specimen that is sterile and will never have the responsibility of caring for others, a specimen that is even-tempered with a narrow and predictable range of expression, all of these are considered 'enhanced'... 'improved'. Better 'performance' is just that; the ability to produce a better result, it does not mean a person with greater skills. It may mean a specimen with narrower skills and the ability to repeat a task.

So while the current ethical debate is about whether or not we should upload computers into our brains, and how human we will be when that happens. There is something happening to us on the nano scale right now; what it is exactly is unknown to us.

Morgellons

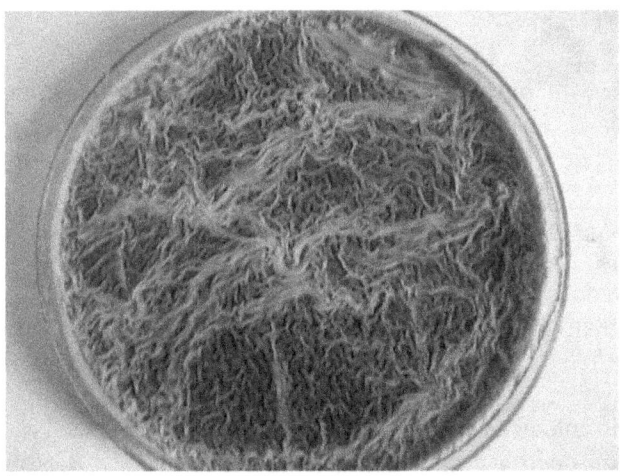

As attempts by lay-people to communicate with scientists about Morgellons disease materials are going nowhere. There is a blackout on this subject, it's victims are dismissed as having a psychological problem; otherwise known as *'delusional parasitosis'*. The presence of patented creations in our bodies, gives rise to intellectual property issues. We know what has been done to small farmers; into whose fields the winds have brought genetically engineered strains. They are sued by the powerful agricultural companies who own the patents.

Canadian farmer Percy Schmeiser was sued by Monsanto in 1998 for patent infringement. "*Monsanto will trespass on a farmer's property, take seed samples -- looking for their GMO seeds -- and then sue the farmer for his land, his crop and all he owns.*" - PoliticalNews.com

Will the day come when *we* are subjected to the jurisdiction of corporations whose patented materials... we are carrying in *our* bodies? It will not matter how it got there, the fault will simply be yours for having it in your possession. This is a forced partnership between us and them, this is how we will be eternally owned by them, this is how they can push our biology from Homo Sapien to Homo Evolutis without our having a say in it.

For now, engineered technology in all living things is a secret, but one day we may be charged with unlawful possession of something that has become a part of us, that we cannot get rid of.

The nature of biology is to adapt. As more unnatural elements enter our bodies, if we cannot reject them, we will find ways to accommodate them. You could call it *'Invasion of the Body Snatchers'*, meets *'Sleeping with the Enemy'*. In fact the original movie *'Invasion of the Body Snatchers'* (1956) contains some interesting lines.

> *"Your new bodies are taking you over - cell for cell, atom for atom ... and you'll be born into an untroubled world."*

> *"Don't fight it, Miles. It's no use."*

> *"Their bodies were now hosts harboring an alien form of life - a cosmic form."*

Metabiology (*beyond* biology; as metaphysics means *beyond* physics) was a term coined by the famous Jonas Salk (of the Salk Polio vaccine), it describes a form of biological prospecting, exploiting genetics using chemistry, physics, and radiation for commercial and other goals. The 1940's and 1950's gave us the birth of radiation biology; the techniques of which were used to decipher the mysteries of heredity, genes, an immunity. Geneticists mixed chemistry and physics, while using X-rays and UV light to irradiate plants, fungi, and fruit flies to see how mutagens altered amino acids and enzymes to form a new biochemistry. All of this continues today.

We are living walking laboratories for powerful science in a society of increasing control. We are being altered. The future being spoken of is happening right now.

Metabiology continues today. We are living laboratories for science. We are being altered.

"Nanobots will infuse all the matter that are around us with information. Rocks, trees, everything will become these intelligent computers. At that point, you can expand out into the rest of the Universe. We will be sending, basically nanotechnology infused with artificial intelligence; swarms of

those will go out in the Universe and basically find other matter and energy that we can then harness to expand the overall intelligence of our human-machine civilization. The Universe will 'wake up', it will become intelligent and that will multiply our intelligence trillions of trillions of times fold. You know we can't really fully contemplate, and that's really the main reason that this is called 'the singularity', but regardless of what you call it, it will be the Universe 'waking up'. So does God exist? Well, I would say not... yet. " - Ray Kurzweil

I am sure you have heard that the military and their black budget projects are decades in advance of conventional technology; have some groups already reached the singularity and are implementing it for the rest of us?

YES! Don't people care about breathing the chemtrails? They WANT this conversion.. this convergence. They want the singularity. I couldn't tell you whether they have attained it or not, but it is what is desired for humanity and all living things. As you can see Ray Kurzweil say that it will be EVERYWHERE. The entire universe is going to be filled with material that responds to external controls. Intelligence is simply that in their definition.

Where is this coming from? We have to become meta-biological experts ourselves, beyond our own biology. Are there frequencies that we are capable of generating through our consciousness, this approaches the realm of metaphysics. We need to put together conclusive studies, I am just like you, I saw different things, and approached those things with a certain perspective and showed you that perspective within this book. Connecting the dots. We are stuck here on the third dimension, reacting and responding that we are creatures of intention, of creation. If we can remember that, and wake that up in ourselves... we have a chance.

Morgellons is not acknowledged as a clinical condition other than a psychological problem, and in fact I think some large organization declared it moot and not pertinent.

Remember the Eugenicists society that were elites in the 1900's, that wanted to weed out all races save for Anglo Saxon. Leaders in California even influenced the Nazi party ... elites making plans. Eugenics was changed to Transhumanism. The post-human will have no gender, non reproductive, improved in the sense that it will be tailored to perform better and accomplish specific tasks. Think of cows on a dairy farm. The improved human is not going to be some kind of *'X-Men'* superhero, it will not be Einstein, it will not be anyone even fancy. When you have technological control of all living things, you can confer attributes to some, and you can degrade and devolve others. You and I are going to be the goblins, not the elites.

How can we transcend this? Learn more at **EvolutisBook.com** and the following websites: **AboutTheSky.com - MorgellonsExposed.com - MoregellonsResearchGroup.com - WayPastHuman.com - Carnicom.com** and **CarnicomInstitute.org**

Extra special thanks to Sophia Smallstrom for all her work, please support her and Clifford Carnicom and all others trying to spread awareness and truth.

20% of all proceeds from this book are donated to Morgellons research.

11 OUR FUTURE

What corporate-motivated "science" has in mind for the future of humanity is far different from the dreamy utopian landscape that's been portrayed by the mainstream media. To hear the corporate-run media tell it, science is always "good" for humanity. Scientific achievements are always called "advances" and not "setbacks," even though many of them have proven to be disastrous for humanity (atomic bombs, for example, or GMOs).

While pure science is, indeed, a necessary component of any civilization which seeks to expand its understanding of the universe, what we see dominating the landscape today isn't pure science but corporate-driven "science" that only seeks to accelerate corporate profits, not human understanding. And with that corporate-slanted science comes a whole new era of truly terrifying technologies that we may soon see become reality in our world.

Future technologies that might be used TODAY to strip away your freedoms and enslave you to the corporate globalist masters, all under the label of "science."

Remember those transgenic pigs... well organ harvesting from genetically modified, patented pigs is very close to happening. Need a replacement heart or lung? No worries! Monsanto will grow you a new one using a genetically modified, trans-species pig (patent pending) that was raised on GMO animal feed and subjected to organ harvesting while it was still alive in order to keep the organs "fresh."

Your government-approved, Medicare-funded transplant will be handled by one of the top U.S. hospitals, which are, even today, deeply engaged in black market organ trafficking and illegal transplantations.

"Behavioral vaccines" are just over the horizon (if not already being covertly administered) that rewire your brain to eliminate dissent. Disobedience is a disease! And the "cure" for disobedience (or Oppositional Defiance Disorder, as they call it) will be a new "vaccine" that biologically rewires your brain to make you more socially acceptable to the controllers. It will be called a "behavioral vaccine" even though, in reality, it's just a chemical lobotomy. This technology will be a cornerstone of the global police state, which will have no tolerance for independent thinking or critical thought of any kind, especially against the state.

You can bet that big brother will not want to miss out on centralized, remote monitoring of all your health statistics and vital signs by the police state. Do you honestly think your medical records are really private? Think again: Even now, the U.S. government maintains a secret centralized bank of blood taken from children at birth. In the near future, citizens will be implanted with biometric monitoring chips that relay information back to the government about your pulse, respiration, and the presence of either illegal drugs or legalized pharmaceuticals (which are often the very same chemicals as illegal drugs, just re-branded as a medication).

These chips will be used by the government to enforce people taking their medications. They will also be used to locate and arrest those who smoke a little pot or take addictive substances without a prescription.

But most importantly, these chips will be used to monitor nutritional levels and make sure no one attains a high level of vitamin D, for example, which promotes clear thinking and strong cognitive function. Under scientific dictatorship, the sheeple must be kept in a state of chronic nutritional deficiency in order to be easily controlled. This will all be sold to the public as a way for the government to monitor their "safety" because, the government will claim, "*Too much vitamin D can be dangerous!*" So they will set the upper safety limits to the lower threshold of cognitive awakening, making sure that everyone remains in a mental stupor as they live out their state-run lives.

The total secrecy of all food ingredients, sources and places of origin are so close... I can *taste* it. As the food industry is increasingly invaded by junk science (GMOs, anyone?), efforts will increase to hide all the chemical ingredients in food products and rename dangerous-sounding chemicals into nice-sounding chemicals.

The Corn Refiners Association is already trying to rename "High Fructose Corn Syrup" to "corn sugar." Aspartame is now going to be called "AminoSweet," and MSG has been renamed things like "yeast extract" or "Torula yeast powder."

But it's going to get far worse as fraudulent science accelerates food industry deceptions. Expect to see preservatives like "sodium benzoate" renamed as things like, *"Freshiness crystals."* Or *"artificial colors"* might be described as *"Fortified with pretty colors."*

Above all, the food industry wants to hide where its foods come from, how they are made, and what's in them, because all three of those categories are bad news for your health.

Get ready for the complete criminalization of home-produced foods and medicines, forcing total reliance on factory food production. Speaking of food, corrupt "scientists" will soon insist that growing your own food is extremely dangerous because you might grow e.coli in your garden (while they grow it in their lab)! With such absurd justifications, home gardening will be completely outlawed in many towns, and those who try to secretly grow tomatoes will be arrested and imprisoned as if they were heroin smugglers.

The idea of all this is to make the population completely dependent on centralized factory food production, in the same way the population is currently dependent on centralized electricity and centralized fossil fuels. This will all be justified with the help of "scientists" who claim that factory-produced food is safer for you because it's all pasteurized, irradiated and fumigated.

We may have already seen the unleashing of a global bioweapon pandemic through seasonal flu shots. Whereas vaccines were once intended to prevent disease, they are now being increasingly weaponized and engineered to spread disease, which is why most of the people who get the flu each winter are the very same people who routinely take flu shots.

In the near future, as the globalists decide the world population has reached its upper tolerable limit, a live *"population control"* virus will be engineered right into the vaccines, followed by an aggressive vaccine push that even offers to pay people to receive flu shots. (Get a flu shot, earn $25!)

The whole scheme, of course, is nothing more than a population control measure designed to eliminate all the lower-IQ people on the planet who are stupid enough to allow themselves to be injected with biological weapons packaged and sold as vaccines. Effectively, it's really a eugenics program that the globalists believe will save the human race from the rise of stupidity (no matter what the cost in human suffering).

Lets not forget about the destruction and extinction of the family unit. Just think about how alone kids of the future will be and how separate and perhaps even without a 'support system' as total government control over your reproduction and the genetic code of your "offspring" becomes the 'norm'.

Copulating with the person of your choice and producing your own "random" offspring will no longer be allowed under the scientific police state. Reproduction must be carefully controlled through licensing and regulation to make sure that no unexpected results occur.

Before having children, parents will need to apply to the government for permission to reproduce, at which point they will be genetically and cognitively profiled, then granted a reproduction classification status that must be strictly followed to avoid imprisonment.

People who show rebellious tendencies and speak out against the state will be denied reproduction "privileges." Only the most obedient, white-skinned, do-gooder mind slaves will be granted reproduction privileges, and they will gladly copulate and raise yet more babies to be sacrificed to the state as the next generation of mind slaves.

And personally, I can't wait for the wireless brain implants that can be remotely activated by law enforcement to make entire crowds of people passive. Bet you never saw a protest group scatter so fast... that is if anyone has the motivation to even leave their Skinner box to go protest in the first place.

The future of "science" involves all sorts of electronics implanted into the human body. One of the most convenient ones will be the "pacification chip" that will be forced upon citizens along with "money chips" that they use to pay for everything (cash will be outlawed, and using cash will be seen as a terrorist activity), and probably both at the same time.

The pacification chip can be remotely activated by the government through cell tower bursts -- or through hand-held units issued to police and law enforcement commanders -- to instantly pacify large crowds of protesters or rioters. Are the students protesting about free speech again? Activate the pacification chip, and they'll all lay down on the lawn and daydream for a while.

Are revolutionaries marching on the capitol and trying to overthrow the government? Activate the pacification chip, and your tyrannical dictatorship is safe!

Such chips may also be used to "excite" the brain at times when it is also politically useful. For example, when another terrorist attack is staged on U.S. soil, the "excitation chips" can be activated across the population to get people riled up and calling for war! (And that's the whole point of false flag attacks, of course.)

The genetic engineering and breeding of obedient super soldiers will eventually be the commonplace. Why use humans to fight when you can just use machines? In the far future, battlefield soldiers will actually be humanoid-shaped robots equipped with firearms and body armor. Think "Terminator" model T-1000. That's still a ways off, of course, given the incredible complexity of mobile power, robotic actuation technologies, vision recognition systems and artificial intelligence.

In the mean time, the most powerful nations (and private interests) of the world will pour R&D money into growing genetically modified super soldiers who are secretly birthed, raised and trained to be as robotic as possible. These super soldiers will be genetically engineered with peak performance attributes (high blood oxygenation, large body frames, etc.) combined with small brains that can only process enough information to follow orders but never question them.

They will also be outfitted with numerous electronic implants, making them more cyborg than human. They will have vision implants attached to their retinas, for example, GPS chips wired to their brains, comm equipment wired into their ears, and built-in pain medication dispensers that flood their bodies with stimulant chemicals so they can keep fighting even after an arm gets blown off, for example.

The electromagnetic activation of metals and nano-crystals injected into you through vaccines and sprayed on you via aerosols will eventually become obvious. In addition to vaccines being used to spread infectious disease, they can also be used to inject humans with nano-crystals that are sized and tuned to resonate at certain frequencies, much like a radio crystal tunes in to a specific radio band.

Such nano-crystals may lie dormant in the bodies of the general public for years or even decades, but at some point the government can take over the radio towers with an "emergency" national transmission that broadcasts an activation signal at precisely the right wavelength to excite the nano-crystals already in peoples' bodies. The results could be anything from mass insanity to massive outbreaks of violence (rioting, etc.) or just tens of millions of people instantly dropping dead. Any of those outcomes could then be exploited by the government to sell a cover story of a "terrorist attack" that requires even more government control over the population.

It could all be done in the name of "science".

Remember, these possible future technologies that exemplify the abuse of science to empower tyrannical governments and corrupt industries are very possible, and are actually in the works as you read this. Will you let this happen? Hopefully, the examples in this chapter have not come true yet, but several are well on their way to become reality in just the next few years.

Real science has an important role to play in any society, but science should serve the interests of the People, not the self-serving controllers who run globalist corporations and national governments.

When science is used to dominate and enslave people rather than setting them free, it is a violation of one of the most fundamental truths throughout the universe: only through freedom (the freedom of ideas, freedom of questioning, freedom of discussion) can true understanding of our universe be achieved.

Awareness is the cure.

ABOUT THE AUTHOR

Trent Goodbaudy has a background in Aviation Maintenance Technology, Computer Science, Computer Information Systems, Programming, Design, Manufacturing, Prototyping, and currently a student of Administrative Law, Sacred Geometry, Hermetic Wisdom, Spirituality, the true history of our planet, and eventually all the secrets of the universe.

Trent is driven to write his books out of a passion for helping others, and he believes that awareness and knowing exactly who you are and who you are not is the most empowering concept one can learn in life.